MOLECULAR GAS DYNAMICS

BY

G. A. BIRD

LAWRENCE HARGRAVE PROFESSOR OF AERONAUTICAL ENGINEERING
UNIVERSITY OF SYDNEY

CLARENDON PRESS · OXFORD
1976

Oxford University Press, Ely House, London W.1

GLASGOW NEW YORK TORONTO MELBOURNE WELLINGTON
CAPE TOWN IBADAN NAIROBI DAR ES SALAAM LUSAKA ADDIS ABABA
DELHI BOMBAY CALCUTTA MADRAS KARACHI LAHORE DACCA
KUALA LUMPUR SINGAPORE HONG KONG TOKYO

ISBN 0 19 856120 2

PRINTED IN GREAT BRITAIN BY
J. W. ARROWSMITH LTD., BRISTOL, ENGLAND

PREFACE

THIS book is intended primarily for scientists and engineers who wish to analyse practical nonlinear gas flows on the molecular level.

The background material for the analysis is drawn largely from the classical kinetic theory of gases. This is developed from first principles and, unlike conventional treatments, it is based on the point of view that a typical element of gas may not be in thermal equilibrium and generally forms part of a flow field with a non-uniform mass motion. Also, since the formulation of molecular gas flow problems need not necessarily be in terms of the velocity distribution function, the macroscopic gas quantities are initially defined as direct averages of the microscopic molecular quantities. The introduction of the various forms of the velocity distribution function is delayed until they are required for the formulation of the Boltzmann equation. The discussion of this equation leads naturally to the equilibrium state. This state, together with the slightly non-equilibrium state that is described by the Chapman–Enskog theory, provides the key reference quantities. These quantities, which are presented for both simple gases and gas mixtures, are supplemented by a discussion of the gas–surface boundary condition. The aforementioned elementary results are sufficient for the treatment of the collisionless or free molecule flows that are dealt with in Chapter 5.

The material in the first five chapters provides the foundation for the treatment of transition regime flows. This material is supplemented by a set of exercises at the end of each chapter and could provide the basis of a graduate course on molecular gas dynamics. The assumed level of knowledge in mathematics and physics is consistent with the minimum requirements for an engineering or science degree. It is only in the later chapters that it is also assumed that readers have been exposed to an introductory course in continuum gas dynamics.

The Chapman–Enskog theory for the transport properties is not presented in detail because a number of full descriptions are readily available in standard texts. Similarly, in the treatment of transition regime flows, emphasis is placed almost exclusively on nonlinear problems with large perturbations. The reasons for this are that other authors have dealt extensively with small perturbation methods, new problems that are encountered in practice usually involve large perturbations, and one of the major purposes of the book is to provide a detailed description of a numerical method that is better suited to flows with large perturbations. This is the direct simulation Monte Carlo method and, following a general survey of transition regime methods in

Chapter 6, a complete chapter is devoted to an introduction to this method. This includes the description of a representative computer program that is listed in an appendix.

One-dimensional transition regime flows of simple monatomic gases are discussed in Chapter 8, while the final four chapters deal with the progressively more difficult classes of multi-dimensional flows, flows of gas mixtures, polyatomic gas flows, and flows involving chemical reactions. A large number of methods are available for the simple one-dimensional flows, but analysis of the more complex flows becomes increasingly reliant on numerical simulation. The necessary extensions to the direct simulation Monte Carlo method are described in detail and several additional programs are listed in appendices. The chapters on transition regime flow contain a number of new results. Most of these have been obtained from the direct simulation method, but minor extensions have been made to results from other methods and much of the background material is new. The book therefore contains elements of an introductory text on molecular gas dynamics, a handbook on the direct simulation Monte Carlo method, and a research monograph on nonlinear transition regime flows. While this has necessarily led to some compromise, and certainly does not allow an encyclopaedic approach, it is hoped that the various elements have been combined in a way that proves useful to both students and practising engineers and scientists.

ACKNOWLEDGEMENTS

The general arrangement of the introductory material, together with consideration of the possibility of writing a book, had its beginnings in a course given by the author as a Visiting Professor of Aeronautics at the California Institute of Technology in 1969. Thanks are due to Dr. H. P. Liepmann for making this possible. The first drafts of the early chapters were written while Dr. G. S. Springer of the University of Michigan was visiting Sydney and the author is indebted to his valuable suggestions and criticisms.

The development of the direct simulation Monte Carlo method has been supported since its inception by the Air force Office of Scientific Research, United States Air Force, through a series of basic research grants. These have been monitored by Mr. M. Rogers and Mr. P. A. Thurston, whose sustained interest is gratefully acknowledged. Not the least of the benefits of this support is that it has enabled the author to participate in most of the bi-annual International Symposia on rarefied gas dynamics. A number of original results from the method appear in the book; this work has been largely supported by the current grant AFOSR-72-2336. Also, many of the Monte Carlo simulation procedures have come into being as a result of a long period of consultation with TRW Systems Inc. (California). Discussions with Drs. F.

W. Vogenitz and J. E. Broadwell have contributed significantly to these developments.

The typing of the manuscript has been capably handled by Mrs. V. Robinson and Mrs. S. Frost. Drs. D. I. Pullin and A. Chatwani have provided some valuable criticisms of the manuscript. Finally, Mr. T. Tran Cong checked a number of chapters and suggested many detailed improvements to the presentation.

Sydney, Australia *G.A.B.*
June, 1975

CONTENTS

LIST OF SYMBOLS

Note: Some symbols, including superscripts and subscripts, that have application only to single problems in Chapter 5 and in Chapters 7–12 have been omitted. This applies also to local changes in the meaning of some listed symbols. FORTRAN variables in the demonstration programs have not been listed.

a	speed of sound
a_c	energy accommodation coefficient, eqn (4.62)
A	a locally specified constant or parameter
$A_2(\eta)$	a definite integral for inverse power law molecules, eqn (4.54)
b	miss distance impact parameter in a binary collision
B	a locally specified constant or parameter
c_p, c_V	specific heats at constant pressure and volume, respectively
c'_m	most probable thermal speed, $c'_m = 1/\beta$
c'_s	root-mean-square thermal speed
$\boldsymbol{c}_0$	stream, mean, or mass velocity, $\boldsymbol{c}_0 = \bar{\boldsymbol{c}}$ for simple gas, eqn (1.37) for gas mixture
$\boldsymbol{c}'$	thermal, peculiar, or random molecular velocity, $\boldsymbol{c}' = \boldsymbol{c} - \boldsymbol{c}_0$
$\boldsymbol{c}_m$	centre of mass velocity of collision pair
$\boldsymbol{c}_r$	relative velocity between two molecules
$\boldsymbol{c}''_p$	velocity defined for species p by $\boldsymbol{c}''_p = \boldsymbol{c}_p - \bar{\boldsymbol{c}}_p$
$\mathrm{d}\boldsymbol{c}$	volume element in velocity space, $\mathrm{d}\boldsymbol{c} = \mathrm{d}u\,\mathrm{d}v\,\mathrm{d}w$
C	a locally specified constant
$\boldsymbol{C}_p$	diffusion velocity of species p, $\boldsymbol{C}_p = \bar{\boldsymbol{c}}_p - \boldsymbol{c}_0$
d	effective or nominal molecular diameter
e_{tr}	specific energy associated with translational modes, eqn (1.21)
e_{int}	specific energy associated with internal modes
$\boldsymbol{e}$	unit vector
E_a	activation energy
E_c	reference energy in collision; total energy in collision
E_t	translational relative energy in collision
f	normalized velocity distribution function in velocity space, eqn (3.1)
f_0	Maxwellian or equilibrium velocity distribution function, eqn (3.46)
f_c	distribution function for molecular speed (subscripts are similarly used to designate the distribution functions for other speeds and velocities)
F	the magnitude of a force; cumulative distribution function
$F^{(N)}$	N particle distribution function, eqn (3.7)
$\boldsymbol{F}$	intermolecular force; external force per unit mass
$\mathscr{F}$	velocity distribution function in phase space, eqn (3.5)

h	Planck's constant, $h = 6{\cdot}6256 \times 10^{-34}$ J s; a linear dimension
H	Boltzmann's H-function, eqn (3.40)
I	molecular moment of inertia
$k(T)$	rate coefficient
k	Boltzmann constant, $k = \mathscr{R}/\mathscr{N} = mR = 1{\cdot}3805 \times 10^{-23}$ J K^{-1}
K	coefficient of heat conduction
(Kn)	Knudsen number, $(Kn) = \lambda/L$
l	linear dimension
l_1	direction cosine with the x-axis
L	linear dimension
m	mass of a single molecule
m_1	direction cosine with the y-axis
m_r	reduced mass, eqn (2.7)
(Ma)	flow Mach number, $(Ma) = c_0/a$
$(Ma)_s$	shock Mach number
$\mathscr{M}$	molecular weight
n	number density
n_0	Loschmidt's number, $n_0 = 2{\cdot}68699 \times 10^{19}$ cm^{-3}
n_1	direction cosine with the z axis
N	number of molecules; number flux of molecules
N_c	number of collisions
$\mathscr{N}$	Avogadro's number, $\mathscr{N} = 6.0225 \times 10^{23}$ mol^{-1}
p	pressure, eqn (1.19); normal momentum flux per unit area
p_{xy}	component of $\boldsymbol{P}$ based on the flux of x momentum in the y direction (similarly for other components)
P	probability
$\mathbf{p}$	pressure tensor, eqn (1.17)
(Pr)	Prandtl Number
q_x	component of q in the x direction
$\boldsymbol{q}$	heat flux vector, eqn (1.27)
Q	physical quantity associated with a molecule
Q^a, Q^{aa}	partition functions
r	radius; radial coordinate
$\boldsymbol{r}$	position vector
$\mathrm{d}\boldsymbol{r}$	volume element in physical space, $\mathrm{d}\boldsymbol{r} = \mathrm{d}x\,\mathrm{d}y\,\mathrm{d}z$
R	gas constant, $R = \mathscr{R}/\mathscr{M}$
R_f	random fraction
$\mathscr{R}$	universal gas constant, $\mathscr{R} = 8{\cdot}3143$ J mol^{-1} K^{-1}
s	molecular speed ratio, $s = U\beta$; distance; number of species in a gas mixture; number of square terms
S	entropy; area
t	time
t_f	energy transfer factor in energy sink model

T	thermodynamic temperature
T_{tr}	translational kinetic temperature, eqn (1.23)
T_{int}	kinetic temperature of internal modes, eqn (1.25)
T_{ov}	overall kinetic temperature, eqn (1.26)
u	velocity component in the x direction
U	speed
v	velocity components in the y direction
V	volume in physical space
w	velocity component in the z direction
W	dimensionless impact parameter, $W = b/r$.
W_f	weighting factor
W_0	dimensionless impact parameter for inverse power law molecules, eqn (2.24)
$W_{0,m}$	cut-off value of W_0
x	Cartesian coordinate axis in physical space, dummy variable
y	Cartesian coordinate axis in physical space
z	Cartesian coordinate axis in physical space
α	angle of incidence, angle, degree of dissociation
β	reciprocal of most probable molecular speed in an equilibrium gas, $\beta = (2RT)^{-\frac{1}{2}}$
γ	ratio of specific heats, $\gamma = c_p/c_V$
Γ	mass flux; gamma function
δ	average spacing between molecules, $\delta = n^{-\frac{1}{3}}$
$\Delta[Q]$	collision integral, eqn (3.26)
ε	azimuth angle impact parameter in binary collision; fraction of molecules specularly reflected at a surface; symmetry factor
ε_{int}	internal energy of a single molecule
ε_0	parameter in Chapman–Enskog expansion, eqn (3.50)
ζ	number of internal degrees of freedom
η	exponent of inverse power law molecular force, eqn (2.23); temperature exponent in generalized Arrhenius equation
θ	angular coordinate
θ_A	angle between apse line and relative velocity vector
κ	constant in inverse power law molecular force, eqn (2.23)
λ	mean free path, eqn (1.8)
λ_0	mean free path in equilibrium gas, eqn (4.38)
μ	coefficient of viscosity
ν	collision frequency, eqn (1.6)
ν_0	collision frequency in an equilibrium gas, eqn (4.36)
$\sigma\,d\Omega$	differential cross-section, eqn (2.11)
σ_T	total collision cross-section, eqn (2.14)
σ_R	reactive cross-section
τ	shear stress; mean collision time
τ_{xy}	viscous stress in the plane normal to x and in the y direction

τ	viscous stress tensor, eqn (1.20)
ϕ	intermolecular potential; azimuth angle
Φ	dissipation function, eqn (3.37); parameter in alternate Chapman–Enskog expansion, eqn (3.51)
χ	deflection angle of the relative velocity vector in a collision
ω_m	most probable angular speed of a molecule in an equilibrium gas
ω	angular velocity
Ω	solid angle

Superscripts and subscripts

Note: The subscripts in the above list do not, in general, conform to the following meanings.

*	post-collision value; sonic conditions
1, 2	particular molecules or molecular classes
p, q	particular molecular species
c	continuum value
f	free molecule or collisionless value
i	incident; intial; inward
r	reflected; relative
∞	freestream value
w	value at a surface
x, y, z	components in the x, y, and z directions, respectively

1

THE MOLECULAR MODEL

1.1 Introduction

A GAS flow may be modelled on either a macroscopic or a microscopic level. The macroscopic model regards the gas as a continuum and the description is in terms of the variations of the macroscopic velocity, density, pressure, and temperature with distance and time. On the other hand, the microscopic or molecular model recognizes the particulate structure of a gas as a myriad of discrete molecules and ideally provides information on the position and velocity of every molecule at all times. However, a description in such detail is rarely, if ever, practical and a gas flow is almost invariably described in terms of macroscopic quantities. The two models must therefore be distinguished by the approach through which the description is obtained, rather than by the nature of the description itself. This book is concerned with the microscopic or molecular approach and the first question which must be answered is whether this approach can solve problems that could not be solved through the conventional continuum approach.

The macroscopic quantities at any point in a flow may be identified with average values of appropriate molecular quantities; the averages being taken over the molecules in the vicinity of the point. The continuum description is valid as long as the smallest significant volume in the flow contains a sufficient number of molecules to establish meaningful averages. The existence of a formal link between the macroscopic and microscopic quantities means that the equations which express the conservation of mass, momentum, and energy in the flow may be derived from either approach. While this might suggest that neither of the approaches can provide information that is not also accessible to the other, it must be remembered that the conservation equations do not form a determinate set unless the shear stresses and heat flux can be expressed in terms of the other macroscopic quantities. It is the failure to meet this requirement, rather than the breakdown of the continuum description, that places a limit on the range of validity of the continuum equations. More specifically, the Navier–Stokes equations of continuum gas dynamics fail when gradients of the macroscopic variables become so steep that their scale length is of the same order as the average distance travelled by the molecules between collisions, or *mean free path*. A less precise but more convenient parameter is obtained if the scale length of the gradients is replaced by a characteristic dimension of the flow. The ratio of the mean free path λ to the characteristic dimension L defines the *Knudsen number* (Kn) i.e.

$$(Kn) = \lambda/L. \tag{1.1}$$

The necessary condition for the validity of the continuum approach is, therefore, that the Knudsen number be small compared with unity.

There appears to be no alternative to the molecular approach when the Knudsen number is of order unity or higher. A high Knudsen number may result from either a large mean free path or a very small characteristic dimension. The former is usually the case and, since the mean free path in a given gas is inversely proportional to the density, it is a consequence of a very low density. Hence the title 'rarefied gas dynamics' that is very frequently applied to the subject of this book. Typical applications in this category include internal flows in vacuum systems and aerodynymics in the outer atmosphere. However, it must be kept in mind that the alternative requirement of a small characteristic dimension can be met at any density. For example, the molecular approach is required for the study of the forces on sufficiently small particles suspended in the atmosphere, the internal structure of shock waves, and the propagation of sound at extremely high frequencies.

Having defined the conditions in which the molecular approach must be used, we must also consider whether there are any circumstances in which it is to be preferred over the continuum approach when both provide valid formulations. This possibility has generally been completely dismissed on account of the overwhelming difficulties associated with formal analytical solutions of the conventional microscopic equations. However, even with the continuum formulation, exact or approximate analytical solutions can only be obtained for comparatively simple flows. Numerical solutions are required for practical problems involving large disturbances and complex boundary conditions. Massive computer calculations are becoming increasingly important in gas dynamics and, for such calculations, the molecular approach will be shown to sometimes offer advantages over the continuum approach.

1.2. The simple dilute gas

The basic quantities associated with the molecular model are the number of molecules per unit volume and the mass, size, and velocity of each molecule. These quantities must be related to the mean free path and collision frequency in order to establish the distance and time scales of the effects due to the collisional interactions among the molecules. Also, since the results from the molecular approach will generally be presented in terms of the macroscopic quantities, we must establish the formal relationships between the macroscopic and microscopic quantities. For reasons of simplicity and clarity, the discussion in this section will be restricted to a gas consisting of a single chemical species in which all the molecules are assumed to have the same structure. Such a gas is called a *simple gas.*

The number of molecules in one mole of gas is a fundamental physical constant called *Avogadro's number* $\mathcal{N}$. Avogadro's law also states that the volume occupied by a mole of any gas at a particular temperature and pressure is a constant. The number of molecules per unit volume, or *number density n*, of a gas therefore depends on the temperature and pressure, but is independent of the composition of the gas. The *mass m* of a single molecule is obtained by dividing the *molecular weight* $\mathcal{M}$ of the gas by Avogadro's number, i.e.

$$m = \mathcal{M}/\mathcal{N} \tag{1.2}$$

The average volume available to a molecule is $1/n$, so the mean molecular spacing δ is given by

$$\delta = n^{-\frac{1}{3}}. \tag{1.3}$$

The ideal molecular structure, from the point of view of the theoretician, would be a hard elastic sphere of diameter d. Then, as shown in Fig. 1.1,

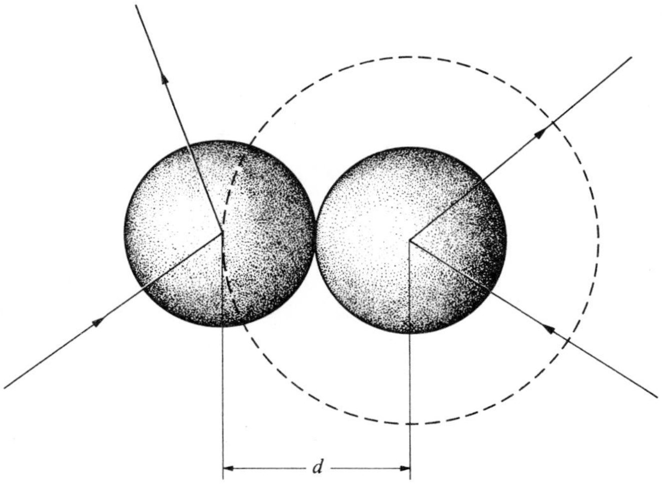

FIG. 1.1. Collision between two hard spheres of diameter d.

two molecules collide if their trajectories are such that the distance between their centres decreases to d. The *total collision cross-section* for these molecules is, therefore,

$$\sigma_{\mathrm{T}} = \pi d^2. \tag{1.4}$$

A real molecule comprises one or more atoms, each consisting of a nucleus surrounded by orbiting electrons. Molecular size is a quantity that cannot be precisely and uniquely defined. However, the effect of a close encounter between molecules can be calculated from a knowledge of the intermolecular force field. The general form of the force field between two neutral molecules is shown in Fig. 1.2 as a function of the distance between them. The force

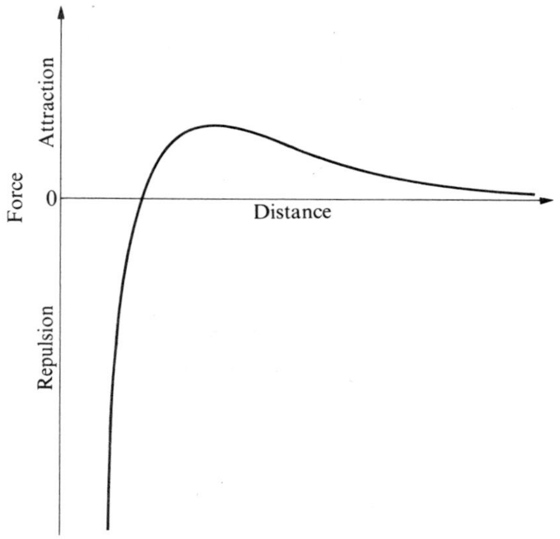

FIG. 1.2. Typical intermolecular force field.

is effectively zero at large distances; it becomes weakly attractive when the molecules are sufficiently close for the interaction to commence, but then decreases again to be very strongly repulsive at short distances. The intermolecular force in a diatomic or polyatomic gas is generally a function of the molecular orientations. Collision dynamics will be dealt with in detail in § 2.2 where procedures for the calculation of collision cross-sections and effective molecular diameters will be discussed. One point that should be mentioned here is that the collision cross-section is generally a function of the relative speed between the collision partners.

The proportion of the space occupied by a gas that actually contains a molecule is of the order of $(d/\delta)^3$. Eqn (1.3) shows that, for sufficiently low densities, the molecular spacing δ is large compared with the effective molecular diameter d. Under these circumstances, only an extremely small proportion of space is occupied by molecules and each molecule will, for the most part, be moving freely in space outside the range of influence of other molecules. Moreover, when it does suffer a collision, it is overwhelmingly likely to be a *binary collision* involving only one other molecule. This situation may be characterized by the condition

$$\delta \gg d, \tag{1.5}$$

and defines a *dilute gas*.

The time scale of the microscopic processes is set by the *mean collision time* which is, by definition, the mean time interval between the successive collisions suffered by a typical molecule. The reciprocal of this quantity is in more common usage and is called the *mean collision rate* or *collision frequency* ν per molecule. In the derivation of an expression for this quantity we will fix our attention on a typical molecule which will be referred to as the *test* molecule. The velocities of the other, or *field*, molecules through which the test molecule moves are distributed in some unspecified manner. Consider those field molecules with velocity between $\boldsymbol{c}$ and $\boldsymbol{c}+\Delta\boldsymbol{c}$. These will be referred to as molecules of class $\boldsymbol{c}$ and their number density is denoted by Δn. If the velocity of the test molecule is $\boldsymbol{c}_t$, the relative velocity between the test molecule and the field molecules of class $\boldsymbol{c}$ is $\boldsymbol{c}_r = \boldsymbol{c}_t - \boldsymbol{c}$. Now choose a frame of reference in which the test molecule moves with velocity $\boldsymbol{c}_r$ and the class $\boldsymbol{c}$ field molecules are stationary. Then, over a time interval Δt much shorter than the mean collision time, the test molecule would collide with any molecule in the cylinder of volume $\sigma_T c_r \Delta t$, as shown in Fig. 1.3. The probability of a collision between the test molecule and a molecule of class $\boldsymbol{c}$ in the time interval Δt is therefore $\Delta n \sigma_T c_r \Delta t$. When collisions do occur, the cylinder swept out by the collision cross section along the trajectory becomes distorted. However, for a dilute gas in which only a small percentage of the trajectory is affected by collisions, the restriction on Δt can be removed and the number of collisions per unit time with a class $\boldsymbol{c}$ molecule is $\Delta n \sigma_T c_r$. The mean collision rate is obtained by summing over all velocity classes and therefore over all values of c_r. That is,

$$\nu = \sum (\Delta n \sigma_T c_r) = n \sum \{(\Delta n / n) \sigma_T c_r\}$$

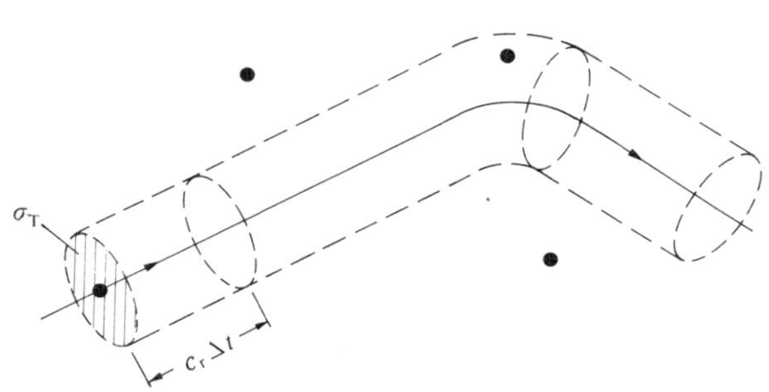

FIG. 1.3. Effective volume swept out by a moving test molecule among stationary field molecules.

and, since $\Delta n/n$ is the fraction of molecules with cross section σ_T and relative velocity $\boldsymbol{c}_r$,

$$\nu = n\overline{\sigma_T c_r}. \tag{1.6}$$

A bar over a quantity or expression denotes the average value over all molecules in the sample. If σ_T is regarded as a constant, the mean collision rate becomes

$$\nu = \sigma_T n\overline{c_r} = \pi d^2 n\overline{c_r}. \tag{1.6a}$$

The total number of collisions per unit time per unit volume of gas is therefore given by

$$N_c = \tfrac{1}{2}n\nu = \tfrac{1}{2}n^2\overline{\sigma_T c_r}. \tag{1.7}$$

The symmetry factor $\frac{1}{2}$ is introduced because each collision involves two molecules.

The *mean free path* is the average distance travelled by a molecule between collisions and is therefore equal to the mean speed $\overline{c'}$ of a molecule divided by the collision frequency i.e.

$$\lambda = \overline{c'}/\nu = \{n(\overline{\sigma_T c_r}/\overline{c'})\}^{-1} \tag{1.8}$$

or, for the constant cross-section case,

$$\lambda = \{(\overline{c_r}/\overline{c'})\pi d^2 n\}^{-1}. \tag{1.8a}$$

The mean free path is defined in a frame of reference moving with the stream velocity of the gas. The prime on the mean molecular speed $\overline{c'}$ denotes that this quantity is measured relative to the stream velocity.

Before moving on to discuss the relationships between the microscopic and macroscopic quantities, we must introduce the concept of *equilibrium* as applied to gases. Consider a volume of gas that is completely isolated from any outside influence. If this gas remains undisturbed for a time that is sufficiently long in comparison with the mean collision time, it may be regarded as being in an equilibrium state. Then, if the number of molecules in the volume is sufficiently large that statistical fluctuations may be neglected, there are no gradients of macroscopic quantities with either distance or time. Also, the fraction of molecules in any velocity class remains constant with time, even though the velocity of an individual molecule changes with each of its collisions. It is obvious that there can be no preferred direction in an equilibrium gas and the velocity distribution must be isotropic. The form of the velocity distribution in an equilibrium gas will be derived in Chapter 3 and its properties will be discussed in Chapter 4. We will find that the continuum approach is adequate for flows that are in equilibrium or depart only slightly from equilibrium.

The first of the macroscopic properties to be discussed is the *density* ρ. This is defined as the mass per unit volume of the gas, and is therefore equal to the product of the number of molecules per unit volume and the mass of a single molecule i.e.

$$\rho = nm. \tag{1.9}$$

Macroscopic quantities such as the density are generally associated with a 'point' in a flow. Values at a 'point' must, in fact, be based on the molecules within a small volume element enclosing the point. An element of volume V contains a number N of molecules and this number is subject to statistical fluctuations about the mean value nV. The probability $P(N)$ of a particular value of N is given by the Poisson distribution

$$P(N) = (nV)^N \exp(-nV)/N!.$$

The standard deviation of this distribution is $(nV)^{\frac{1}{2}}$, corresponding to a fractional deviation from the mean of $(nV)^{-\frac{1}{2}}$. For large values of nV, the distribution is indistinguishable from a normal or Gaussian distribution.† The expected magnitude of the fluctuations is illustrated in Fig. 1.4. This indicates that a generally excessive level of scatter is present unless a macroscopic property is based on a sample nV of the order of 1000. Eqn (1.3) enables nV to be written V/δ^3, where δ is the mean molecular spacing. Therefore, for a meaningful result, the establishment of a macroscopic quantity by averaging over the molecules in a small element of volume requires that the typical dimension $V^{\frac{1}{3}}$ of the element should satisfy the condition

$$V^{\frac{1}{3}} \gg \delta. \tag{1.10}$$

Unless the dimensions of the volume element are small compared with the scale length L of the macroscopic gradients, the macroscopic quantities based on the molecules in the element will depend on the size of the element. For a three-dimensional flow with gradients in all directions, this leads to the requirement that $V^{\frac{1}{3}}$ should be much smaller than L. However, for a one or two-dimensional flow, the volume element may be elongated in the direction or directions with zero gradients. This means that the dimension in the direction of a gradient may be kept small compared with the scale length L and the volume sufficiently large to satisfy the condition (1.10), even though $V^{\frac{1}{3}}$ may be larger than L.

The above discussion deals only with *instantaneous averages* in a single flow or system. Two additional types of average must be taken into account

† If $nV \gg 1$ and $|N - nV|/nV \ll 1$,

$$\frac{(nV)^N}{N!}\exp(-nV) \simeq \frac{1}{(2\pi nV)^{\frac{1}{2}}}\exp\left\{-\frac{(N-nV)^2}{2nV}\right\}.$$

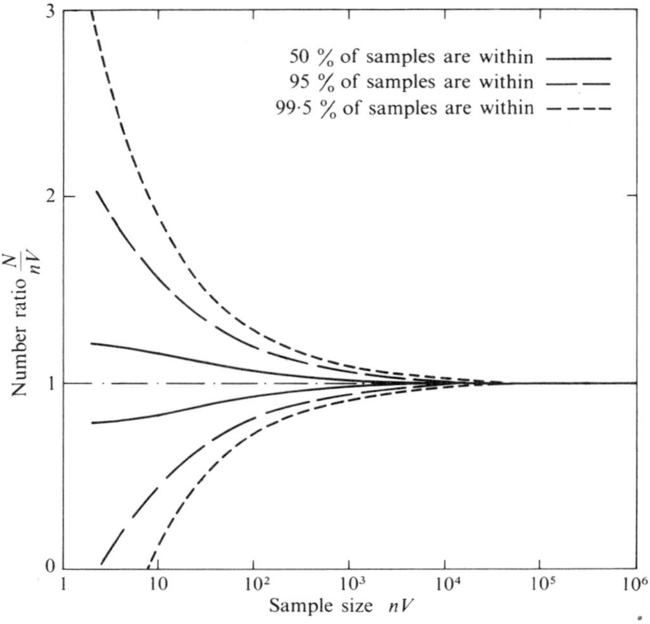

FIG. 1.4. Statistical fluctuations as a function of sample size.

before conclusions are made about the conditions which allow meaningful macroscopic quantities to be defined for the flow. The first is established by summing the appropriate properties of the molecules in the volume over an extended time interval. This is called the *time average* and enables any steady flow to be described in terms of the macroscopic quantities. The second average is an instantaneous average taken over the volume elements in an arbitrarily large ensemble of similar systems. This *ensemble average* can be established wherever an experiment or calculation can be repeated indefinitely. Almost any flow can be described by macroscopic quantities established through time or ensemble averages, although the description is essentially probabilistic and the fluctuations may have to be taken into account. When both time and ensemble averages can be used to describe a flow, the two descriptions can generally be expected to be identical. The molecular motion is then said to be *ergodic*.

The remaining macroscopic quantities of interest are related to the transport of mass, momentum, and energy in the flow as a result of the molecular motion. Before proceeding with the discussion of these quantities, we must establish an important general result for the flux of some quantity Q across a surface element at some location in the gas. This relates the flux to averages taken over the molecules in a volume element at the same location. The

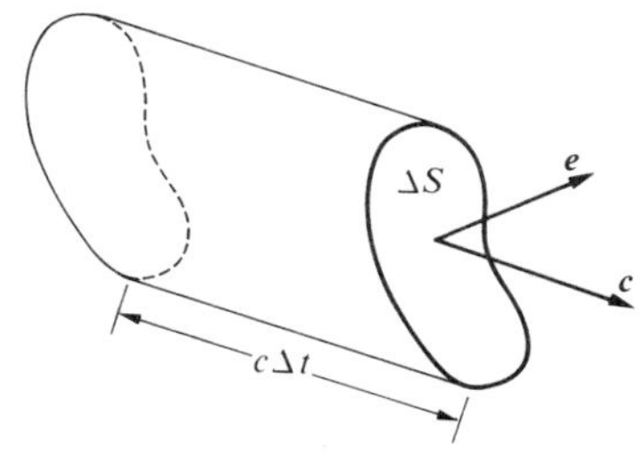

FIG. 1.5. Flux of molecules of class $\boldsymbol{c}$ across a surface element.

surface element has area ΔS and unit normal vector $\boldsymbol{e}$, as shown in Fig. 1.5. Consider the molecules of class $\boldsymbol{c}$ with velocity between $\boldsymbol{c}$ and $\boldsymbol{c}+\Delta\boldsymbol{c}$ and let their number density again be Δn. The molecules of class $\boldsymbol{c}$ that cross ΔS in a short time interval Δt are contained, at the beginning of Δt, within the cylinder projected from ΔS in the direction opposite to $\boldsymbol{c}$ with length $c\Delta t$. The height of this cylinder measured in the direction of $\boldsymbol{e}$, normal to ΔS, is $\boldsymbol{c}\cdot\boldsymbol{e}\Delta t$ and its volume is $\boldsymbol{c}\cdot\boldsymbol{e}\Delta S\Delta t$. The quantity Q is associated with each molecule and is either a constant or a function of $\boldsymbol{c}$. The flux of Q across the surface per unit area per unit time in the direction of $\boldsymbol{e}$ is, therefore, $\Delta nQ\boldsymbol{c}\cdot\boldsymbol{e}$. The total flux is obtained by summing over all velocity classes, and can be written

$$n\sum\left(\frac{\Delta n}{n}Q\boldsymbol{c}\cdot\boldsymbol{e}\right),$$

or

$$n\overline{Q\boldsymbol{c}\cdot\boldsymbol{e}}. \tag{1.11}$$

Note that the expression (1.11) is for the total flux and includes the contributions from molecules crossing the surface element in both the positive and negative $\boldsymbol{e}$ directions. We will sometimes require the flux in one direction only and, without loss of generality, this may be taken as the positive $\boldsymbol{e}$ direction. The required flux is then obtained by considering only those molecules with a positive value of $\boldsymbol{c}\cdot\boldsymbol{e}$. The flux in the positive $\boldsymbol{e}$ direction is, therefore,

$$n(\overline{Q\boldsymbol{c}\cdot\boldsymbol{e}})_{\boldsymbol{c}.\boldsymbol{e}>0}. \tag{1.12}$$

It has been assumed in the derivation of the expressions (1.11) and (1.12) that the quantity Q is transferred across the element ΔS only when the molecule with which it is associated crosses the element. This is exact for the number and mass fluxes, but is not generally true when the momentum flux or energy flux is being considered. This is because the finite range of the intermolecular force field permits an interchange of momentum and energy between molecules that remain on the opposite sides of the element. However, if the gas is dilute with $\delta \gg d$, these effects may be neglected.

The necessity for averages to be based on a sufficiently large sample, unless one resorts to time or ensemble averages, appears to impose a further restriction on the expressions (1.11) and (1.12). The dimensions of the volume element in Fig. 1.5 should be large in comparison with the mean molecular spacing; i.e. $\bar{c}\Delta t$ should be much larger than δ. Also, the derivation of these equations was based on the assumption that Δn is a constant. We have seen that this is the case in an equilibrium gas. However, in a non-equilibrium situation, the number of molecules scattered out of class $\boldsymbol{c}$ as a result of collisions is not equal to the number scattered into the class. The analysis is therefore valid in a non-equilibrium gas only if Δt is much less than the mean collision time $1/\nu$. The gas must therefore be such that $\bar{c}/\nu \gg \delta$ or, from eqn (1.8), $\lambda \gg \delta$. Eqns (1.8) and (1.3) show that

$$\frac{\lambda}{\delta} = \frac{1}{\pi(\overline{c_r}/\overline{c'})}\left(\frac{\delta}{d}\right)^2. \tag{1.13}$$

The ratio $\overline{c_r}/\overline{c'}$ is of order unity and, consequently, $\lambda \gg \delta$ whenever the dilute gas condition ($\delta \gg d$) is satisfied. The ordering $\lambda \gg \delta \gg d$ is sometimes used as the definition of a dilute gas, but eqn (1.13) shows that eqn (1.5) is a sufficient condition.

Since $\boldsymbol{e}$ is a unit vector, the expression (1.11) can be written $(n\overline{Q\boldsymbol{c}})\cdot\boldsymbol{e}$ and therefore defines the *flux vector* for the quantity Q as

$$n\overline{Q\boldsymbol{c}}. \tag{1.14}$$

The flux vector related to the transport of mass is obtained by setting Q equal to the molecular mass, to give,

$$nm\overline{\boldsymbol{c}} \quad \text{or} \quad \rho\overline{\boldsymbol{c}}.$$

The mean molecular velocity $\overline{\boldsymbol{c}}$ defines the macroscopic *stream* or *mean* or *mass velocity* which is denoted by $\boldsymbol{c}_0$ i.e.

$$\boldsymbol{c}_0 = \overline{\boldsymbol{c}}. \tag{1.15}$$

The velocity of a molecule relative to the stream velocity is called the *thermal* or *peculiar* or *random velocity* and is denoted by $\boldsymbol{c}'$ i.e.

$$\boldsymbol{c}' = \boldsymbol{c} - \boldsymbol{c}_0. \tag{1.16}$$

Note that,

$$\overline{\boldsymbol{c}'} = \overline{\boldsymbol{c}} - \boldsymbol{c}_0 = \overline{\boldsymbol{c}} - \overline{\boldsymbol{c}} = 0,$$

so the mean thermal velocity is zero in a simple gas. The remaining macroscopic quantities are defined by averages taken over the peculiar velocities of the molecules. Therefore, in order to simplify the discussion of these quantities, we will now look at the element of gas from a frame of reference moving with the local stream velocity.

The flux vector for the peculiar velocities is $n\overline{Qc'}$ and an expression for the momentum transport by the peculiar or thermal motion is obtained by setting Q equal to $m\boldsymbol{c}'$. Since momentum is a vector quantity, the resulting expression is a tensor with nine cartesian components, called the pressure tensor $\mathbf{p}$, i.e.

$$\mathbf{p} = nm\overline{\boldsymbol{c}'\boldsymbol{c}'} = \rho\overline{\boldsymbol{c}'\boldsymbol{c}'} \tag{1.17}$$

and is best explained in terms of the separate components. Let u', v', and w' be the components of $\boldsymbol{c}'$ in the x, y, and z directions and, as an example, consider the x momentum flux in the y direction. That is, set $Q = mu'$ and choose the surface element in the x, z plane so that the normal is the y direction. This component is

$$p_{xy} = \rho\overline{u'v'},$$

and the complete set is:

$$\begin{aligned} p_{xx} &= \rho\overline{u'^2}, & p_{xy} &= \rho\overline{u'v'}, & p_{xz} &= \rho\overline{u'w'}; \\ p_{yx} &= p_{xy}, & p_{yy} &= \rho\overline{v'^2} & p_{yz} &= \rho\overline{v'w'}; \\ p_{zx} &= p_{xz}, & p_{zy} &= p_{yz} \quad \text{and} & p_{zz} &= \rho\overline{w'^2}. \end{aligned} \tag{1.18}$$

A shorthand way of writing the nine equations (1.18) for the components of $\mathbf{p}$ is

$$p_{ij} = \rho\overline{c_i'c_j'}, \tag{1.18a}$$

where the subscripts i and j each range from 1 to 3. The values 1, 2, and 3 may be identified with the components along the x, y, and z axes, respectively i.e.

$$c_1' \equiv u', \qquad c_2' \equiv v' \quad \text{and} \quad c_3' \equiv w'.$$

The scalar *pressure* p is usually defined as the average of the three normal components of the pressure tensor i.e.

$$p = \tfrac{1}{3}\rho(\overline{u'^2}+\overline{v'^2}+\overline{w'^2}) = \tfrac{1}{3}\rho\overline{c'^2}. \tag{1.19}$$

If the gas is in equilibrium, the three normal components are equal and the pressure is given by the product of the density and the mean value of the square of the peculiar velocity components in any direction. Consider the case in which the gas is bounded by a solid surface and choose the reference direction normal to the surface. If the gas is also in equilibrium with the surface, the molecules reflected from the surface will be indistinguishable from the molecules that would come from an imaginary continuation of the gas on to the other side of the surface. The scalar pressure p may then be identified with the normal force per unit area exerted by the gas on the surface. However, in a non-equilibrium situation, the pressure defined by eqn (1.19) is not identical with the normal force per unit area on an element of solid surface. The latter quantity is usually of greater practical importance and, when there is an ambiguity, the term 'pressure' is usually applied to the normal force per unit area.

The *viscous stress tensor* $\boldsymbol{\tau}$ is defined as the negative of the pressure tensor with the scalar pressure subtracted from the normal components. It is most conveniently represented in the component or subscript notation as

$$\boldsymbol{\tau} \equiv \tau_{ij} = -(\rho\overline{c_i'c_j'} - \delta_{ij}p), \tag{1.20}$$

where δ_{ij} is the Kronecker delta such that

$$\delta_{ij} = 1 \text{ if } i = j, \text{ and } \delta_{ij} = 0 \text{ if } i \neq j.$$

The average kinetic energy associated with the thermal or translational motion of a molecule is $\frac{1}{2}m\overline{c'^2}$, and the specific energy associated with this motion is

$$e_{\text{tr}} = \tfrac{1}{2}\overline{c'^2}. \tag{1.21}$$

This may be combined with eqn (1.19) to give

$$p = \tfrac{2}{3}\rho e_{\text{tr}}$$

which may be compared with the ideal gas equation of state

$$p = \rho RT = nkT. \tag{1.22}$$

Here, k is the Boltzmann constant which is related to the universal gas constant $\mathscr{R}$ by $k = \mathscr{R}/\mathscr{N}$. Also $m = \mathscr{M}/\mathscr{N}$ and the ordinary gas constant $R = \mathscr{R}/\mathscr{M}$, so that $k = mR$. The *thermodynamic temperature* T is essentially an equilibrium gas property, but the above comparison shows that the ideal gas equation of state will apply to a dilute gas, even in a non-equilibrium situation, for a *translational kinetic temperature* T_{tr} defined by

$$\tfrac{3}{2}RT_{\text{tr}} = e_{\text{tr}} = \tfrac{1}{2}\overline{c'^2},$$

or (1.23)

$$\tfrac{3}{2}kT_{\text{tr}} = \tfrac{1}{2}m\overline{c'^2} = \tfrac{1}{2}m(\overline{u'^2}+\overline{v'^2}+\overline{w'^2}).$$

Note that separate translational kinetic temperatures may be defined for each component. For example,

$$kT_{\text{tr}_x} = m\overline{u'^2} = p_{xx}/n \tag{1.24}$$

and the departure of these component temperatures from T_{tr} provides a convenient measure of the degree of translational non-equilibrium in a gas.

Monatomic molecules may be assumed to possess translation energy only. Therefore, in a monatomic gas, the translational temperature may be regarded simply as the *temperature* T. However, diatomic and polyatomic molecules also possess internal energy associated with the rotational and vibrational energy modes. Since three degrees of freedom are associated with the translational mode, a temperature T_{int} for the internal modes may be defined consistently with the translational temperature (see eqn 1.23) by

$$\tfrac{1}{2}\zeta R T_{int} = e_{int}. \tag{1.25}$$

Here, ζ is the number of internal degrees of freedom and e_{int} is the specific energy associated with the internal modes. The principle of equipartition of energy states that the translational and internal temperatures must be equal in an equilibrium gas; the common values may then be identified with the thermodynamic temperature of the gas. An *overall kinetic temperature* T_{ov} may be defined for a non-equilibrium gas as the weighted mean of the translational and internal temperatures, i.e.

$$T_{ov} = (3T_{tr} + \zeta T_{int})/(3+\zeta), \tag{1.26}$$

Note that the ideal gas equation of state does not apply to this temperature in a non-equilibrium situation.

Finally, the heat flux vector $\boldsymbol{q}$ is obtained by setting Q equal to the molecular energy $\frac{1}{2}mc'^2 + \varepsilon_{int}$, i.e.

$$\boldsymbol{q} = \tfrac{1}{2}\rho\overline{c'^2\boldsymbol{c}'} + n\overline{\varepsilon_{int}\boldsymbol{c}'} \tag{1.27}$$

Here, ε_{int} is the internal energy of a single molecule and is related to e_{int} by $e_{int} = \varepsilon_{int}/m$.

The component of heat flux in the x direction is written

$$q_x = \tfrac{1}{2}\rho\overline{c'^2u'} + n\overline{\varepsilon_{int}u'}. \tag{1.27a}$$

1.3. Extension to gas mixtures

Consider a gas mixture consisting of a total of s separate chemical species. Values pertaining to a particular species will be denoted by the subscripts p or q, each of which may range from 1 to s. The overall number density n is obviously equal to the sum of the number densities of all the individual species i.e.

$$n = \sum_{p=1}^{s} n_p. \tag{1.28}$$

Consider a collision between a molecule of species p and one of species q. The effective diameters are d_p and d_q, and the requirement for a collision is that the distance between their centres decreases to $\frac{1}{2}(d_p+d_q)$. The total collision cross-section is therefore

$$\sigma_{\mathrm{T}_{pq}} = \tfrac{1}{4}\pi(d_p+d_q)^2 = \pi d_{pq}^2,$$

where (1.29)

$$d_{pq} = \tfrac{1}{2}(d_p+d_q).$$

A molecule of species p may be taken as the test molecule and the molecules of species q as the field molecules in the analysis leading to eqn (1.6). The mean collision rate for a species p molecule with a species q molecules is, therefore,

$$\nu_{pq} = n_q\overline{\sigma_{\mathrm{T}_{pq}}c_{\mathrm{r}_{pq}}}, \tag{1.30}$$

where $c_{\mathrm{r}_{pq}}$ is the relative speed between the two molecules. If $\sigma_{\mathrm{T}_{pq}}$ is regarded as a constant, this becomes

$$\nu_{pq} = \pi d_{pq}^2 n_q \overline{c_{\mathrm{r}_{pq}}}. \tag{1.30a}$$

The mean collision rate for species p molecules is obtained by summing over all types of collision partner i.e.

$$\nu_p = \sum_{q=1}^{s} (n_q\overline{\sigma_{\mathrm{T}_{pq}}c_{\mathrm{r}_{pq}}}). \tag{1.31}$$

A mean collision rate per molecule for the mixture may be defined by averaging over all types of test particle to give

$$\nu = \sum_{p=1}^{s} \{(n_p/n)\nu_p\}. \tag{1.32}$$

The number of collisions per unit time per unit volume between species p molecules and species q molecules is

$$n_p\nu_{pq}$$

or

$$n_p n_q\overline{\sigma_{pq}c_{\mathrm{r}_{pq}}}. \tag{1.33}$$

Note that, when $p = q$, this expression counts each collision twice over. Eqn (1.33) may be summed over all species q molecules to give the total number of collision per unit volume per unit time that involve species p molecules as $n_p\nu_p$. This, in turn may be summed over all type p molecules to determine the total number of collisions N_{c} per unit time per unit volume.

Since all collisions are then counted twice over, the symmetry factor of $\frac{1}{2}$ is introduced to give

$$N_c = \tfrac{1}{2} \sum_{p=1}^{s} (n_p \nu_p) = \tfrac{1}{2} n \nu,$$

as would be expected from eqn (1.7).

The mean free path for a species p molecule is equal to its mean thermal speed divided by its collision frequency i.e.

$$\lambda_p = \overline{c'_p}/\nu_p = \left\{ \sum_{q=1}^{s} (n_q \overline{\sigma_{\mathrm{T}_{pq}} c_{r_{pq}}} / \overline{c'_p}) \right\}^{-1} \tag{1.34}$$

and the mean free path for the mixture is

$$\lambda = \sum_{p=1}^{s} \{(n_p/n)\lambda_p\}. \tag{1.35}$$

The macroscopic density is equal to the sum of the individual species densities and can be written

$$\rho = \sum_{p=1}^{s} (m_p n_p). \tag{1.36}$$

The stream velocity $\boldsymbol{c}_0$ has been defined for the simple gas such that $\rho \boldsymbol{c}_0$ is equal to the flux vector for molecular mass. A similar procedure for the gas mixture yields

$$\boldsymbol{c}_0 = \frac{1}{\rho} \sum_{p=1}^{s} (n_p m_p \overline{\boldsymbol{c}_p}). \tag{1.37}$$

For the mixture, $\boldsymbol{c}_0$ is called the *mass average velocity*. This is not the mean velocity, but a weighted mean with the weighting of each molecule being proportional to its mass. The momentum of a gas is as if all the molecules move with $\boldsymbol{c}_0$ and it is this velocity that appears in the conservation equations.

The peculiar or thermal velocity $\boldsymbol{c}'$ of each molecule is again measured relative to $\boldsymbol{c}_0$, i.e.

$$\boldsymbol{c}' = \boldsymbol{c} - \boldsymbol{c}_0,$$

and the mean thermal velocity of species p is

$$\overline{\boldsymbol{c}'_p} = \overline{\boldsymbol{c}_p} - \boldsymbol{c}_0. \tag{1.38}$$

Therefore, the mean thermal velocity of a particular species is equal to its mean velocity relative to the mass average velocity. This quantity is called the *diffusion velocity* and is denoted by $\boldsymbol{C}_p$. Therefore,

$$\boldsymbol{C}_p \equiv \overline{\boldsymbol{c}'_p} = \overline{\boldsymbol{c}_p} - \boldsymbol{c}_0. \tag{1.39}$$

The definitions of the pressure tensor, the scalar pressure, the viscous stress tensor, the translational kinetic temperature, and the heat flux vector are rendered valid for a gas mixture simply by including the molecular mass within the averaging process. For example, the scalar pressure is

$$p = \tfrac{1}{3}\sum_{p=1}^{s}(n_p m_p \overline{c_p'^2})$$

$$= \tfrac{1}{3}n\sum_{p=1}^{s}\{(n_p/n)m_p\overline{c_p'^2}\}$$

or

$$p = \tfrac{1}{3}n\overline{mc'^2}. \tag{1.40}$$

The translational kinetic temperature is now defined by

$$\tfrac{3}{2}kT_{\text{tr}} = \tfrac{1}{2}\overline{mc'^2} \tag{1.41}$$

and, in an equilibrium gas, T_{tr} can again be regarded simply as the temperature with the mixture obeying the ideal gas equation of state.

In a non-equilibrium situation, it is convenient to define separate species temperatures to serve as a quantitative measure of the degree of non-equilibrium between the species. Eqn (1.41) can be written

$$\tfrac{3}{2}kT_{\text{tr}} = \tfrac{1}{2}\sum_{p=1}^{s}\{(n_p/n)m_p\overline{c_p'^2}\},$$

and the obvious definition of a species translational kinetic temperature is

$$\tfrac{3}{2}kT_{\text{tr}_p} = \tfrac{1}{2}m_p\overline{c_p'^2}. \tag{1.42}$$

However, the thermal velocity $\boldsymbol{c}'_p$ is measured relative to the mass average velocity $\boldsymbol{c}_0$ which involves all species. It is therefore desirable to relate the kinetic temperature defined by eqn (1.42) to one defined in a similar manner, but based on the *single species thermal velocity* $\boldsymbol{c}''_p$ measured relative to the average velocity $\overline{\boldsymbol{c}}_p$ of the species, i.e.

$$\boldsymbol{c}''_p = \boldsymbol{c}_p - \overline{\boldsymbol{c}}_p. \tag{1.43}$$

This definition, together with eqns (1.38) and (1.39), enables eqn (1.42) to be written

$$\tfrac{3}{2}kT_{\text{tr}_p} = \tfrac{1}{2}\overline{mc_p''^2} + \tfrac{1}{2}m_p C_p^2. \tag{1.44}$$

The kinetic temperature of species p, as defined in eqn (1.42), therefore provides a measure of the sum of the 'single species thermal energy' and the 'kinetic energy of diffusion' of this species.

As in the simple gas case, rotational and vibrational kinetic temperatures may be defined for a gas consisting of diatomic or polyatomic molecules.

Eqn (1.25) applies to the mixture as long as ζ is interpreted as the mean number of internal degrees of freedom. The single species translational temperature defined by eqn (1.42) may again be subdivided into separate temperatures based on $\overline{u_p'^2}$, $\overline{v_p'^2}$, and $\overline{w_p'^2}$, but these must be assessed in conjunction with the component version of eqn (1.44) in order to determine the degree of translational non-equilibrium within each species. All the kinetic temperatures become equal and equivalent to the thermodynamic temperature when the gas is in equilibrium.

1.4. Typical magnitudes

The accepted value of Avogadro's number for the number of molecules in one mole of gas is $6{\cdot}0225 \times 10^{23}$. The number density of any gas under *standard conditions* of 1 atm (101 325 Pa) and 0 °C is then

$$n = 2{\cdot}68699 \times 10^{25}\ \mathrm{m}^{-3}. \tag{1.45}$$

This quantity is called the *standard number density* and the number of molecules in 1 cm^3 is called *Loschmidt's number* n_0. The value of the number density n under other conditions is easily calculated since it is directly proportional to the macroscopic density.

The discussion of the values of the other microscopic quantities will be based on air. For the purpose of this discussion, air will be assumed to be a simple gas of identical 'average air' molecules. The molecular weight of air is approximately 28·97 and eqn (1.2) then gives $4{\cdot}8 \times 10^{-26}$ kg† as the mass of a single molecule. An effective molecular diameter is conveniently obtained by substituting the measured value of the coefficient of viscosity into the theoretical result for hard-sphere molecules (Chapman and Cowling 1952). For air at 0 °C, this yields $d = 3{\cdot}7 \times 10^{-10}$ m. The mean molecular spacing under standard conditions is obtained from n_0 through eqn (1.3) as $\delta_0 = 3{\cdot}3 \times 10^{-9}$ m. Air under standard conditions therefore satisfies the dilute gas condition that $\delta \gg d$, although not by a wide margin. The third distance of interest is the molecular mean free path defined by eqn (1.8). It will be shown in Chapter 4 that the value of $\overline{c_r}/\overline{c'}$ in an equilibrium gas is $\sqrt{2}$ and, therefore,

$$\lambda = (\sqrt{2}\pi d^2 n)^{-1}. \tag{1.47}$$

The above values of d and n_0 then give $6{\cdot}1 \times 10^{-8}$ m for the mean free path in equilibrium air under standard conditions.

† SI units have been used throughout and, since the 'SI mole' is effectively a 'gram mole' rather than a 'kilogram mole', a factor of 10^{-3} has been applied to the right hand side of eqn (1.2).

The mean square molecular speed is given from eqns (1.22) and (1.23) as

$$\overline{c'^2} = 3p/\rho = 3RT = 3(\mathscr{R}/\mathscr{M})T = 3kT/m. \tag{1.48}$$

The root mean square molecular speed $(\overline{c'^2})^{\frac{1}{2}}$ therefore differs from the speed of sound in the gas (which is given by $a^2 = \gamma RT$) only by a constant of order unity. Using the previously quoted value of the molecular weight of air, we obtain 485 m s^{-1} for the root mean square speed of an air molecule at 0 °C. For an equilibrium gas, we will find that the mean thermal speed $\overline{c'}$ is related to the root mean square speed by $\overline{c'} = (8/3\pi)^{\frac{1}{2}} (\overline{c'^2})^{\frac{1}{2}}$. We have already noted that $\overline{c_r} = \sqrt{2}\overline{c'}$ in an equilibrium gas, and eqn (1.6) then gives $7{\cdot}3 \times 10^9$ s^{-1} for the mean collision rate of an air molecule under standard conditions. The total collision rate per unit volume then follows from eqn (1.7) as $9{\cdot}8 \times 10^{34}$ m^{-3} s^{-1}.

We are now in a position to define the limits of validity of the dilute gas assumption, the continuum approach, and the continuum description. The limits are conveniently expressed as functions of the gas density ρ and the characteristic dimension of the flow L. The density may be normalized by the density ρ_0 under standard conditions, but L is best retained as a dimensional quantity. With a double logarithmic plot of L versus ρ/ρ_0, the three limits may be defined by straight lines, as shown in Fig. 1.6.

The dilute gas assumption requires that $\delta/d \gg 1$, and $\delta/d = 7$ has been chosen as the limit. Since both δ and d are independent of L, this is a vertical line and a scale for δ/d has been set along the upper edge of Fig. 1.6.

The validity of the continuum approach has been identified with the validity of the Navier–Stokes equations. This requires that the Knudsen number $(Kn) = \lambda/L$ should be small compared with unity, and $(Kn) = 0{\cdot}1$ has been chosen as the limit. This is a reasonable assumption as long as L is related to the scale length of the appropriate macroscopic variable. For example, in the flow past an aerodynamic body, the Knudsen number based on a typical dimension of the body can be used to predict the general breakdown of the continuum approach. However, the Navier–Stokes equations will fail in the boundary layer when the Knudsen number based on the boundary layer thickness is approximately 0·1. This will usually occur when the Knudsen number based on the typical dimension is very much smaller. Also, at still higher densities, it might be necessary to modify the boundary conditions of the Navier–Stokes equations in order to allow finite slip velocities at solid surfaces.

It was stated in § 1.1 that the continuum description is valid as long as the smallest significant volume in the flow contains a sufficient number of molecules to establish meaningful averages. This is strictly correct, in that a gas can scarcely be described as a continuum when the particular structure of the gas manifests itself through significant fluctuations of the macroscopic quantities. However, we have also seen that ensemble and time averages

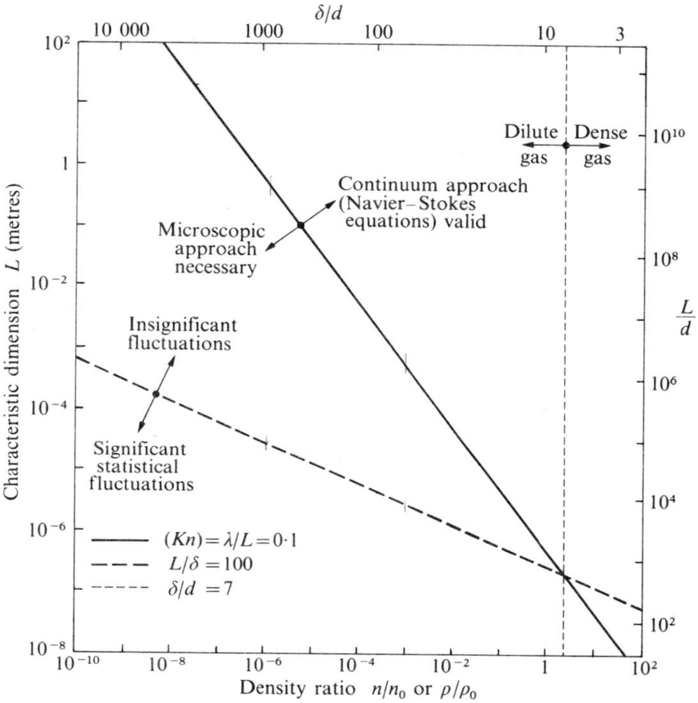

FIG. 1.6. Effective limits of major approximations. ($d = 3{\cdot}7 \times 10^{-10}$ m).

permit a probabilistic description of the flow in terms of the macroscopic or continuum quantities, despite the fluctuations. For this reason, we will identify the third limit with the 'onset of significant fluctuations' rather than the 'breakdown of the continuum description'. The criterion for this limit has been chosen as $L/\delta = 100$. This corresponds to a value of L that is ten times as large as the side of a cubic element containing 1000 molecules. Fig. 1.4 shows that the density in such an element would involve statistical fluctuations with a standard deviation of approximately three per cent.

A feature of Fig. 1.6 is that the lines describing the limits intersect at a single point. While the precise definitions of the limits is arbitrary, this result would not be substantially altered by any reasonable modification of those chosen here. One consequence is that the limit for the validity of the Navier–Stokes equations always lies between the dilute gas limit and the limit for significant fluctuations. Therefore, as the density and/or the characteristic dimension is reduced in a dilute gas, the Navier–Stokes equations become invalid before the level of statistical fluctuations becomes significant. The more rarefied the gas, the wider the margins between these limits. On the other hand, a significant level of fluctuation may be present in a dense gas even when the Navier–Stokes equations are valid. For example, the theory of Brownian motion is partially based on these equations although the phenonemon is, itself, a manifestation of significant statistical fluctuations.

It was noted earlier that the existence of directions with zero macroscopic gradients can affect the way in which the fluctuation conditions are interpreted. As far as Fig. 1.6 is concerned, this means that it may be necessary in some cases to define different characteristic dimensions for the two limits that involve this quantity. For example, consider the problem of one-dimensional heat transfer between parallel plates. The characteristic dimension for the Knudsen number is obviously the spacing between the plates, but the dimension for the ratio L/δ should be the cube root of the product of the plate area and the plate spacing. This is because it is the ratio of this quantity to the mean molecular spacing that determines whether there are significant fluctuations in the heat flux between the plates.

The definition of the molecular model has implicitly assumed the validity of the classical description of a gas. One possible difficulty is that the quantum mechanics description does not allow the precise specification of molecular positions and momenta. The Heisenberg uncertainty principle states that the product of the uncertainty $|\Delta r|$ in molecular position and the uncertainty $|\Delta mc|$ in momentum is of the order of Planck's constant i.e.

$$|\Delta r|\,|\Delta mc| \sim h. \tag{1.49}$$

For the classical model to be appropriate, the uncertainties in position and momentum should be much less than the mean molecular spacing and the mean magnitude of the molecular momentum, respectively. This requires that

$$\delta m\bar{c} \gg h \tag{1.50}$$

and means that δ should be much larger than the mean de Broglie wavelength $(h/m\bar{c})$ of the molecules. The condition for gross diffraction effects to be absent in collisions is that the molecular diameter d should be large compared with the de Broglie wavelength. The second condition is the more restrictive in a dilute gas, but, at this stage, our primary concern is with the general molecular description rather than with the details of the collision processes.

A measure of the molecular speed may be obtained from eqn (1.48) and, using eqn (1.3), eqn (1.50) becomes

$$\frac{(3mkT)^{\frac{1}{2}}}{n^{\frac{1}{3}}h} \gg 1. \tag{1·51}$$

A further condition for the validity of the classical model is that the number of available quantum states should be very large in comparison with the number of molecules. The standard result for the ratio of the number of available states to the number of molecules is, to within a numerical factor of order unity, equal to the cube of the expression on the left hand side of equation (1.51). It is, therefore, easily satisfied whenever eqn (1.51) is satisfied.

The substitution of the 'average air' magnitudes into the expression on the left hand side of eqn (1.51) leads to a value of 117·4. The dilute gas assumption would break down well before the density increased sufficiently to violate eqn (1.51). The only circumstance in which a quantum description becomes necessary for the general molecular model is when a light gas is at an extremely low temperature.

Representative data on the molecules of gases other than air is presented in Appendix A. The only property that affects Fig. 1.6 is the effective molecular diameter d. Table A1 indicates that there is only a factor of three variation between the largest and smallest effective diameters and those for oxygen and nitrogen fall about the middle of the range.

Exercises

1.1. Show that all molecules in a macroscopically homogeneous, simple, dilute gas have the same collision probability if the total collision cross-section σ_T is inversely proportional to the relative speed c_r between colliding molecules.

1.2. Fix attention on one molecule of the gas described in Exercise 1.1. Let P_t be the probability of it not suffering another collision for a time t following a collision. Show that

$$P_t = e^{-\nu t},$$

and that the probability of the molecule travelling a distance L between collisions is

$$P_L = e^{-L/\lambda}.$$

To what extent can this analysis be applied to the more general case in which the collision probability is velocity dependent?

1.3. Consider a hard-sphere gas in which all molecules have the same speed and all molecular directions are equally possible. Show that the mean free path is given by

$$\lambda = \{(4/3)\pi d^2 n\}^{-1}.$$

1.4. A 'two-dimensional gas' in which the motion of the molecules is confined to a plane, rather than to three-dimensional space, has sometimes been postulated as a device to simplify the analysis of kinetic theory problems. Show that the result of Exercise 1.3 in a two-dimensional gas is

$$\lambda = (4d^2 n)^{-1}.$$

1.5. Show that, for a simple gas,

$$\overline{c'^2} = \overline{c^2} - \bar{c}^2 = \overline{c^2} - c_0^2.$$

1.6. A small volume of argon contains just ten molecules. An instantaneous measure of the velocity components yields the following results:

Molecule	u	v	w (m s^{-1})
1	320	−423	−268
2	−463	197	299
3	−217	−254	−108
4	346	291	−212
5	−510	320	−508
6	243	−217	375
7	478	365	251
8	−172	285	−366
9	523	336	−481
10	−387	256	178

Calculate the stream velocity components and the kinetic temperature of the gas in this volume. (Note that the result of Exercise 1.5 can be used to avoid the computation of the thermal velocity components.) (Ans: $u_0 = 16{\cdot}1$, $v_0 = 115{\cdot}6$, $w_0 = -84{\cdot}0$ m s^{-1}, $T = 522{\cdot}5$ K).

1.7. Consider a cubic element of gas with the length of the sides equal to the mean free path. Show that the number of molecules in the element is inversely proportional to the square of the density, and that the total collision rate in the element is inversely proportional to the density.

1.8. Show that the total (i.e. based on the molecular velocities $\boldsymbol{c}$ rather than the thermal velocities $\boldsymbol{c}'$) x momentum flux in the y direction in a simple gas

$$\rho\overline{uv} = \rho\overline{u'v'} + \rho u_0 v_0 = p_{xy} + \rho u_0 v_0.$$

Similarly, show that the total translational molecular energy per unit volume (or translational energy density) in a simple gas

$$\tfrac{1}{2}\rho\overline{c^2} = \tfrac{1}{2}\rho\overline{c'^2} + \tfrac{1}{2}\rho c_0^2 = \tfrac{3}{2}\rho R T_{\text{tr}} + \tfrac{1}{2}\rho c_0^2.$$

1.9. The results of Exercise 1.8 show that there is no coupling between the thermal and stream velocities in the momentum flux and energy density in a simple gas. Derive the following result to show that there is coupling as far as the heat flux is concerned;

$$\tfrac{1}{2}\rho\overline{uc^2} = \tfrac{1}{2}\rho\overline{u'c'^2} + \rho u_0(\tfrac{1}{2}\overline{c'^2} + \tfrac{1}{2}c_0^2) + \rho u_0\overline{u'^2} + \rho v_0\overline{u'v'} + \rho w_0\overline{u'w'}.$$

1.10. Verify eqn (1.44).

1.11. Show that the translational energy density in a gas mixture

$$\tfrac{1}{2}\sum_{p=1}^{s} n_p m_p \overline{c_p^2} = \tfrac{3}{2}nkT_{\text{tr}} + \tfrac{1}{2}\rho c_0^2$$

and, noting that the gas constant in a gas mixture is $R = nk/\rho$, this result is identical with the simple gas result of Exercise 1.8.

1.12. The mean number of molecules in a volume V of a macroscopically uniform gas of number density n is nV. Show that, for $nV \gg 1$, the probability of the actual number being within $A\surd(nV)$ of the mean is $\operatorname{erf}(A/\surd 2)$.

2

BINARY ELASTIC COLLISIONS

2.1. Momentum and energy considerations

WE have seen that intermolecular collisions in dilute gases are overwhelmingly likely to be binary collisions involving just two molecules. An elastic collision is defined as one in which there is no interchange of translational and internal energy. The pre-collision velocities of the two collision partners in a typical binary collision may be denoted by $\boldsymbol{c}_1$ and $\boldsymbol{c}_2$. Given the physical properties of the molecules and the orientation of the trajectories, our task is to determine the post-collision velocities $\boldsymbol{c}_1^*$ and $\boldsymbol{c}_2^*$.

Linear momentum and energy must be conserved in the collision. This requires

$$m_1\boldsymbol{c}_1+m_2\boldsymbol{c}_2 = m_1\boldsymbol{c}_1^*+m_2\boldsymbol{c}_2^* = (m_1+m_2)\boldsymbol{c}_m \tag{2.1}$$

and

$$m_1c_1^2+m_2c_2^2 = m_1c_1^{*2}+m_2c_2^{*2} \tag{2.2}$$

Here, m_1 and m_2 are the masses of the two molecules and $\boldsymbol{c}_m$ is the velocity of the centre of mass of the pair of molecules. Eqn (2.1) shows that this centre of mass velocity is unaffected by the collision. The pre-collision and post-collision values of the relative velocity between the molecules may be defined by

$$\boldsymbol{c}_r = \boldsymbol{c}_1-\boldsymbol{c}_2$$

and (2.3)

$$\boldsymbol{c}_r^* = \boldsymbol{c}_1^*-\boldsymbol{c}_2^*.$$

Eqns (2.1) and (2.3) may be combined to give

$$\boldsymbol{c}_1 = \boldsymbol{c}_m+\frac{m_2}{m_1+m_2}\boldsymbol{c}_r$$

and (2.4)

$$\boldsymbol{c}_2 = \boldsymbol{c}_m-\frac{m_1}{m_1+m_2}\boldsymbol{c}_r.$$

The pre-collision velocities relative to the centre of mass are $\boldsymbol{c}_1-\boldsymbol{c}_m$ and $\boldsymbol{c}_2-\boldsymbol{c}_m$. Eqn (2.4) shows that these velocities are parallel in this frame of reference and, if the molecules are point centres of force, the force between them remains in the plane containing the two velocities. The collision is therefore planar in the centre of mass frame. The post-collision velocities may similarly be obtained from eqns (2.1) and (2.3) as

$$\mathbf{c}_1^* = \mathbf{c}_m + \frac{m_2}{m_1+m_2}\mathbf{c}_r^*$$

and (2.5)

$$\mathbf{c}_2^* = \mathbf{c}_m - \frac{m_1}{m_1+m_2}\mathbf{c}_r^*.$$

This shows that the post-collision velocities are also parallel in the centre of mass frame. The conservation of angular momentum requires that the projected distance between post-collision velocities be equal to the projected distance b between the pre-collision velocities.

Eqs (2.4) and (2.5) show that

$$m_1c_1^2 + m_2c_2^2 = (m_1+m_2)c_m^2 + m_rc_r^2$$

and (2.6)

$$m_1c_1^{*2} + m_2c_2^{*2} = (m_1+m_2)c_m^2 + m_rc_r^{*2},$$

where

$$m_r = \frac{m_1m_2}{m_1+m_2} \tag{2.7}$$

is called the *reduced mass*. A comparison of eqn (2.6) with the energy equation (2.2) shows that the magnitude of the relative velocity is unchanged by the collision, i.e.

$$c_r^* = c_r. \tag{2.8}$$

Since both $\mathbf{c}_m$ and c_r may be calculated from the pre-collision velocities, the determination of the post-collision velocities reduces to the calculation of the change in direction χ of the relative velocity vector.

If $\boldsymbol{F}$ is the force between two spherically symmetric point centre of force molecules and $\boldsymbol{r}_1$, $\boldsymbol{r}_2$ are their position vectors, the equations of motion of the molecules are

$$m_1\ddot{\boldsymbol{r}}_1 = \boldsymbol{F}$$

and (2.9)

$$m_2\ddot{\boldsymbol{r}}_2 = -\boldsymbol{F}.$$

Hence,

$$m_1 m_2(\ddot{\boldsymbol{r}}_1 - \ddot{\boldsymbol{r}}_2) = (m_1 + m_2)\boldsymbol{F}$$

or, if the relative velocity vector is denoted by $\boldsymbol{r}$,

$$m_r \ddot{\boldsymbol{r}} = \boldsymbol{F}. \tag{2.10}$$

The motion of the molecule of mass m_1 relative to the molecule of mass m_2 is therefore equivalent to the motion of a molecule of mass m_r relative to a fixed centre of force.

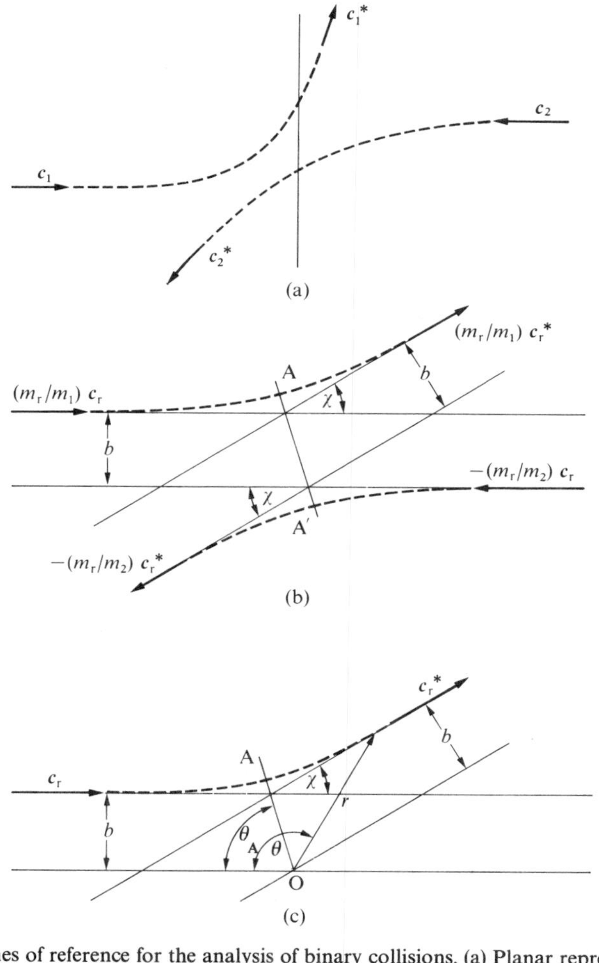

FIG. 2.1. Frames of reference for the analysis of binary collisions. (a) Planar representation of a collision in the laboratory frame of reference. (b) Binary collision in the centre of mass frame of reference. (c) Interaction of the reduced mass particle with a fixed scattering centre.

The above results are summarized in Fig. 2.1. The transformation from the laboratory to the centre of mass coordinate system brings the trajectories into a single plane and illustrates their symmetry about the *apse line* AA′. This symmetry reflects the symmetry of the equations with respect to the pre-collision and post-collision velocities. A further consequence of this symmetry becomes apparent if we consider a collision between two molecules of velocity $\boldsymbol{c}_1^*$ and $\boldsymbol{c}_2^*$ and such that the separation of their undisturbed trajectories in the centre of mass frame of reference is again equal to b. This collision results in post-collision velocities of $\boldsymbol{c}_1$ and $\boldsymbol{c}_2$, and is called the *inverse* of the original or *direct* collision. The trajectories of the direct and inverse encounters are illustrated in Fig. 2.2.

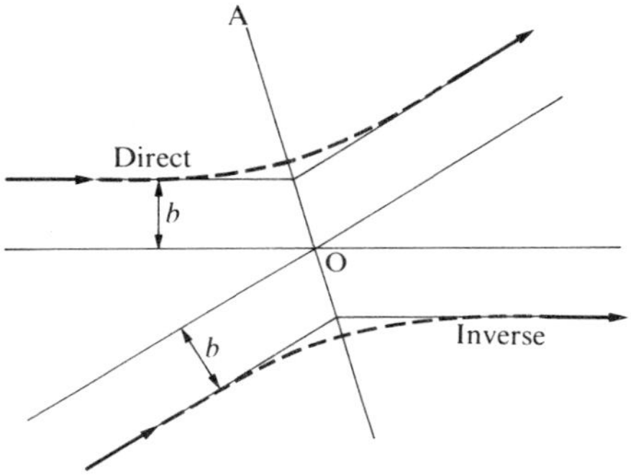

FIG. 2.2. Representation of the direct and inverse encounters of a reduced mass particle with a fixed scattering centre.

2.2. Impact parameters and collision cross-sections

Apart from the translational velocities of the two collision partners, just two *impact parameters* are required to completely specify a binary elastic collision between spherically symmetric molecules. The first is the distance of closest approach b of the undisturbed trajectories in the centre of mass frame of reference. The plane in which the trajectories lie in the centre of mass frame is called the collision plane, and the second impact parameter is chosen as the angle between the plane and some reference plane. As shown in Fig. 2.3, the line of intersection of the collision and reference planes is parallel to $\boldsymbol{c}_r$. If we now consider the plane normal to $\boldsymbol{c}_r$ and containing 0, the *differential cross-section* $\sigma\, d\Omega$ for the collision specified by the impact parameters b and ε is defined by

$$\sigma\, d\Omega = b\, db\, d\varepsilon. \tag{2.11}$$

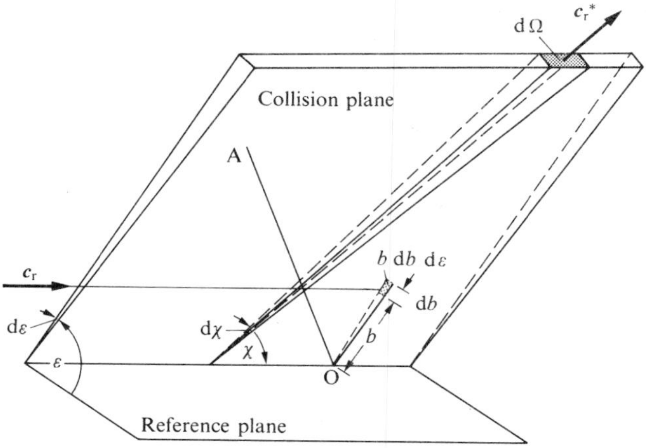

FIG. 2.3. Illustration of the impact parameters.

where $d\Omega$ is the unit solid angle about the vector $\boldsymbol{c}_r^*$. Fig. 2.3 shows that

$$d\Omega = \sin\chi\, d\chi\, d\varepsilon \tag{2.12}$$

so that

$$\sigma = \frac{b}{\sin\chi}\left|\frac{db}{d\chi}\right|. \tag{2.13}$$

Finally, the *total collision cross-section* σ_T is defined by

$$\sigma_T = \int_0^{4\pi} \sigma\, d\Omega = 2\pi \int_0^{\pi} \sigma \sin\chi\, d\chi. \tag{2.14}$$

We will find that this integral diverges for the more realistic molecular models and that it is necessary to introduce effective or nominal cross-sections.

2.3. Collision dynamics

With reference to the polar co-ordinates r, θ defined in Fig. 2.1(c), the angular momentum and energy of the particle of reduced mass m_r within the force field about 0 may be equated to the limiting values of these quantities as $r \to \infty$. The angular momentum equation is simply

$$r^2\dot{\theta} = \text{const} = bc_r. \tag{2.15}$$

The energy within the force field is the sum of the kinetic and potential energies and, since the force tends to zero at large distances, this may be equated to the asymptotic kinetic energy, i.e.

$$\tfrac{1}{2}m_r(\dot{r}^2+r^2\dot{\theta}^2)+\phi = \text{const} = \tfrac{1}{2}m_r c_r^2. \tag{2.16}$$

Here, ϕ is the intermolecular potential which is related to the spherically symmetric force $\boldsymbol{F}$ between the molecules by

$$\phi = \int_r^\infty F\,dr,$$

or (2.17)

$$F = -d\phi/dr.$$

Time may be eliminated from eqns (2.15) and (2.16) to give the equation of the orbit as

$$\left(\frac{dr}{d\theta}\right)^2 = \frac{r^4}{b^2}-r^2-\frac{\phi r^4}{\tfrac{1}{2}m_r c_r^2 b^2}$$

Introducing the dimensionless coordinate

$$W = b/r, \tag{2.18}$$

this becomes

$$(dW/d\theta)^2 = 1-W^2-\phi/(\tfrac{1}{2}m_r c_r^2),$$

so that

$$\theta = \int_0^W \{1-W^2-\phi/(\tfrac{1}{2}m_r c_r^2)\}^{-\frac{1}{2}}\,dW.$$

Now, at the intersection of the orbit with the apse line OA,

$$\theta = \theta_A$$

and

$$dW/d\theta = 0.$$

Therefore

$$\theta_A = \int_0^{W_1} \{1-W^2-\phi/(\tfrac{1}{2}m_r c_r^2)\}^{-\frac{1}{2}}\,dW, \tag{2.19}$$

where W_1 is the positive root of the equation

$$1-W^2-\phi/(\tfrac{1}{2}m_r c_r^2) = 0. \tag{2.20}$$

Finally, the deflection angle of the relative velocity vector is

$$\chi = \pi-2\theta_A. \tag{2.21}$$

The above determination of an expression for the deflection angle χ constitutes the key step in the analysis of dynamics of a binary elastic collision. However, as noted at the beginning of the chapter, the overall objective is to determine the post-collision velocities $\boldsymbol{c}_1^*$ and $\boldsymbol{c}_2^*$. In a typical calculation, the components and magnitudes of $\boldsymbol{c}_m$ and $\boldsymbol{c}_r$ would be obtained from the pre-collision velocities of the collision partners through eqns (2.1) and (2.3). The specification of the impact parameter b then allows the deflection angle χ to be calculated from eqns (2.18) to (2.21). The components of $\boldsymbol{c}_r^*$ are then required in order for the post-collision velocities to be calculated through eqn (2.5). To this end, a set of Cartesian coordinates x', y', and z' may be introduced with x' in the direction of $\boldsymbol{c}_r$. The components of $\boldsymbol{c}_r^*$ along these axes are then $c_r \cos\chi$, $c_r \sin\chi\cos\varepsilon$, and $c_r \sin\chi \sin\varepsilon$. The direction cosines of x' are u_r/c_r, v_r/c_r, and w_r/c_r. Since the orientation of the reference plane is arbitrary, the y' axis may be chosen such that it is normal to the x axis. The direction cosines of y' are then 0, $w_r(v_r^2+w_r^2)^{-\frac{1}{2}}$, and $-v_r(v_r^2+w_r^2)^{-\frac{1}{2}}$ and those of z' follow as $(v_r^2+w_r^2)^{\frac{1}{2}}/c_r$, $-u_r v_r(v_r^2+w_r^2)^{-\frac{1}{2}}/c_r$, and $-u_r w_r(v_r^2+ +w_r^2)^{-\frac{1}{2}}/c_r$. The required expressions for the components of c_r^* in the original x, y, and z coordinates are, therefore,

$$u_r^* = \cos\chi u_r + \sin\chi \sin\varepsilon (v_r^2+w_r^2)^{\frac{1}{2}},$$

$$v_r^* = \cos\chi v_r + \sin\chi (c_r w_r \cos\varepsilon - u_r v_r \sin\varepsilon)/(v_r^2+w_r^2)^{\frac{1}{2}}, \tag{2.22}$$

and

$$w_r^* = \cos\chi w_r - \sin\chi (c_r v_r \cos\varepsilon + u_r w_r \sin\varepsilon)/(v_r^2+w_r^2)^{\frac{1}{2}}.$$

2.4. Particular models

As noted in Chapter 1, the force between two molecules is strongly repulsive at short distances and weakly attractive at larger distances. Considerations of analytical tractability indicate the use of the simplest acceptable model. A model may be said to be acceptable for a particular application if it leads to sufficiently accurate correlations between theory and experiment.

The most useful model completely neglects the attractive component of the force field. This is called either the *inverse power law model* or the *point centre of repulsion model* and is defined by

$$F = \kappa/r^\eta$$

or (2.23)

$$\phi = \kappa/\{(\eta-1)r^{\eta-1}\}.$$

The ratio of the potential energy to the asymptotic kinetic energy may be written

$$\frac{\phi}{\frac{1}{2}m_r c_r^2} = \frac{2\kappa}{(\eta-1)m_r c_r^2 r^{\eta-1}} = \frac{2}{\eta-1}\left(\frac{W}{W_0}\right)^{\eta-1},$$

where W_0 is a second dimensionless impact parameter defined by

$$W_0 = b\left(\frac{m_r c_r^2}{\kappa}\right)^{1/(\eta-1)}. \tag{2.24}$$

Eqn (2.19) to (2.21) then show that the deflection angle is given by

$$\chi = \pi - 2\int_0^{W_1}\left\{1 - W^2 - \frac{2}{\eta-1}\left(\frac{W}{W_0}\right)^{\eta-1}\right\}^{-\frac{1}{2}} dW \tag{2.25}$$

where W_1 is the positive root of the equation

$$1 - W^2 - \frac{2}{\eta-1}\left(\frac{W}{W_0}\right)^{\eta-1} = 0.$$

Note that, for a given η, χ is a function of the dimensionless impact parameter W_0 only. This single parameter dependence of the deflection angle is the basic reason for the usefulness of the inverse power law model. The differential cross-section is a function of c_r and, for a fixed value of c_r, eqn (2.24) and its derivative may be substituted into eqn (2.11), to give

$$\sigma\, d\Omega = W_0\{\kappa/(m_r c_r^2)\}^{2/(\eta-1)}\, dW_0\, d\varepsilon. \tag{2.26}$$

For any finite value of η, the force field extends to infinity and the integral in eqn (2.14) for the total collision cross-section diverges.

The problem of an unbounded total collision cross-section σ_T is common to most classical models. Although σ_T is infinite, the vast majority of collisions are glancing collisions involving very slight deflections. It can be shown (Vincenti and Kruger 1965, p. 359) that, when quantum effects are taken into account, the uncertainty principle does not allow these collisions to be properly defined. Therefore, when applying the classical models, a finite *cut-off* is a theoretical requirement as well as a practical necessity. Such a cut-off may be based on either the miss distance b or the deflection angle χ. In most cases, the latter is to be preferred and, since χ is a function of W_0 for the inverse power law model, a deflection angle cut-off for this model may be applied through the specification of a maximum value $W_{0,m}$ of W_0. Then, for a fixed value of c_r, the total cross section σ_T may be obtained by integrating eqn (2.26) over all possible values of the impact parameters W_0 and ε, i.e.

$$\sigma_T = \int_0^{2\pi}\int_0^{W_{0,m}} W_0\left(\frac{\kappa}{m_r c_r^2}\right)^{2/(\eta-1)} dW_0\, d\varepsilon$$

or

$$\sigma_{\mathrm{T}} = \pi W_{0,m}^2 \{\kappa/(m_r c_r^2)\}^{2/(\eta-1)}. \tag{2.27}$$

However, since this contains the arbitrary cut-off $W_{0,m}$, it is not suitable for setting the effective collision frequency or mean free path.

The calculation of the transport properties, such as the coefficient of viscosity, will be found to involve the differential rather than the total cross-section. A nominal total cross-section for a particular model may therefore be defined as the cross-section of the hard sphere molecule that matches the calculated value of the coefficient of viscosity for that model. This procedure will be adopted when the inverse power law model is applied to a flow of a particular Knudsen number, and effectively fixes the value of the constant κ. Eqn (2.27) shows that the total cross-section of an inverse power law molecule is inversely proportional to $c_r^{4/(\eta-1)}$.

The nominal total cross-section that is established through the above procedure is not to be confused with the viscosity cross-section σ_μ. This is defined by

$$\sigma_\mu = \int_0^{4\pi} \sin^2 \chi \sigma \, d\Omega$$

or, from eqn (2.12),

$$\sigma_\mu = 2\pi \int_0^{\pi} \sigma \sin^3 \chi \, d\chi \tag{2.28}$$

and its name is derived from the fact that this is a convergent integral that is encountered in the Chapman–Enskog theory for the coefficient of viscosity. Also, it may be seen from Fig. 2.1(c) that the component of post-collision velocity in the direction of the pre-collision velocity in the centre of mass frame of reference is $(1-\cos\chi)c_r$. This leads to the momentum transfer cross-section σ_{M}, defined by,

$$\sigma_{\mathrm{M}} = \int_0^{4\pi} (1-\cos\chi)\sigma \, d\Omega$$

or

$$\sigma_{\mathrm{M}} = 2\pi \int_0^{\pi} \sigma(1-\cos\chi)\sin\chi \, d\chi. \tag{2.29}$$

Both σ_μ and σ_{M} have been used as the basis of alternative procedures for the establishment of effective mean free paths and collision frequencies.

The *hard-sphere model* that was employed in much of the discussion in Chapter 1 may be regarded as the special case of the inverse power law model with $\eta = \infty$. As shown in Fig. 2.4, this force becomes effective at

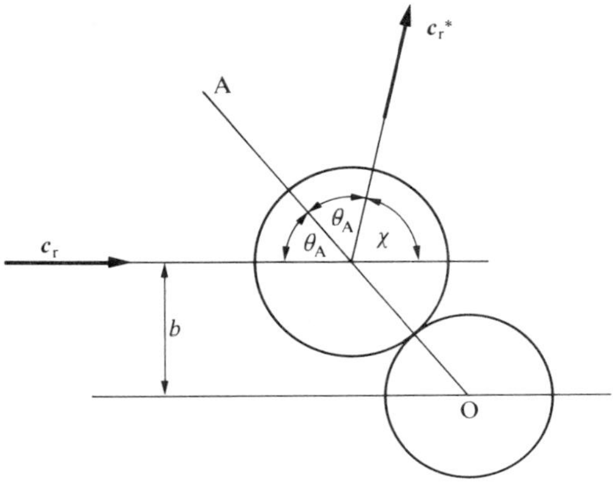

FIG. 2.4. Collision geometry of hard sphere molecules.

$r = \frac{1}{2}(d_1 + d_2) = d_{12}$ and the apse line is the line through the centres of the spheres. Therefore,

$$b = d_{12} \sin \theta_A = d_{12} \cos(\tfrac{1}{2}\chi)$$

and

$$\left|\frac{db}{d\chi}\right| = \tfrac{1}{2} d_{12} \sin(\tfrac{1}{2}\chi),$$

so that eqn (2.13) gives

$$\sigma = d_{12}^2/4. \tag{2.30}$$

This equation shows that σ is independent of χ and that the scattering from hard sphere molecules is isotropic. That is, all directions are equally likely for $\boldsymbol{c}_r^*$, as can be readily seen from purely geometric considerations. The total collision cross-section is, as expected,

$$\sigma_T = \int_0^{4\pi} \sigma \, d\Omega = \pi d_{12}^2. \tag{2.31}$$

It is this finite σ_T that makes the hard-sphere molecule useful in establishing the effective cross-section for other models, as explained above.

The *Maxwell model* is the special case of the inverse power law model with $\eta = 5$. We then have

$$W_1 = W_0^2[\{1 + (2/W_0^4)\}^{\frac{1}{2}} - 1]^{\frac{1}{2}}$$

for the non-dimensional distance of closest approach and eqn (2.25) may be integrated. The result is:

$$\chi = \pi - \frac{2}{\{1+(2/W_0^4)\}^{\frac{1}{4}}} K[\tfrac{1}{2} - \tfrac{1}{2}\{1+(2/W_0^4)\}^{-\frac{1}{2}}], \tag{2.32}$$

where

$$K(\alpha) = \int_0^{\frac{1}{2}\pi} \frac{dy}{(1-\alpha \sin^2 y)^{\frac{1}{2}}}$$

is the complete elliptic integral of the first kind. Eqn (2.26) for the differential cross-section becomes

$$\sigma \, d\Omega = \frac{W_0}{c_r} \left(\frac{\kappa}{m_r} \right)^{\frac{1}{2}} dW_0 \, d\varepsilon. \tag{2.33}$$

We saw in § 1.2 that the collision probability for a pair of molecules is proportional to the product of the cross-section and the relative speed. Eqn (2.33) therefore shows that the collision probability for a particular molecule in a Maxwellian gas is independent of its velocity. This is the major analytical advantage of the Maxwell molecule.

To the extent that a real monatomic molecule is properly represented by the inverse power law model, the effective value of η is generally around ten. The special cases of hard sphere and Maxwell molecules provide useful limiting cases for 'hard' and 'soft' molecules respectively. The effective value of η is generally established through the temperature dependence of the coefficient of viscosity. Theory based on the power law model provides a temperature exponent which is a function of η only. Since the actual exponent can only be regarded as a constant over a limited temperature range, there is sometimes a requirement for a more realistic model incorporating an attractive component.

The *square-well model* adds a uniform attractive potential of finite strength to the hard sphere model, while the *Sutherland model* adds an inverse power law attractive component to the hard-sphere model. A more realistic and probably the best known of the attractive–repulsive models is the *Lennard-Jones potential* which adds an inverse power law attractive component to the inverse power law model. A detailed discussion of these and other models is presented in Hirschfelder, Curtiss, and Bird (1954).

It should be kept in mind that the justification for using any molecular model is that information on the collision cross-sections of the real molecules does not exist, would not affect the results, or renders the analysis intractable. The latter justification becomes increasingly irrelevant as more powerful numerical methods are introduced. Theoretical and experimental information on collision cross-sections is gradually becoming available and Toennies

(1974) has summarized current progress. The use of the simple classical models is most dubious at low temperatures where quantum effects are important and, for some monatomic gases, it is now possible to replace the models by realistic cross-sections.

Exercises

2.1. Verify eqns (2.4) and (2.5).

2.2. Show that, for the special case of $\eta = 2$ (Coulomb molecules),

$$\chi = 2\sin^{-1}\{(1+W_0^2)^{-\frac{1}{2}}\}.$$

2.3. Show that the equation of the trajectory in the centre of mass frame of reference for Maxwell molecules is

$$\theta = \xi^{-\frac{1}{2}}F(\sin^{-1}[W\xi^{\frac{1}{2}}\{1+\tfrac{1}{2}W^2(\xi-1)\}^{-\frac{1}{2}}], \sin^{-1}\{(\tfrac{1}{2}-\tfrac{1}{2}\xi^{-1})^{\frac{1}{2}}\}),$$

where

$$F(\phi, \alpha) = \int_0^{\phi} (1-\sin^2\alpha \sin^2\beta)^{-\frac{1}{2}}\,\mathrm{d}\beta$$

is the elliptic integral of the first kind and

$$\xi = \{1+(2/W_0^4)\}^{\frac{1}{2}}.$$

2.4. Calculate the momentum and viscosity cross-sections for hard sphere and inverse power law molecules.

2.5. Determine the minimum deflection angle that corresponds to a cut-off value of $W_{0,m} = 1{\cdot}5$ for the dimensionless impact parameter W_0 in a gas of (*i*) Maxwell molecules and (*ii*) inverse twelfth-power molecules. (Ans: (*i*) $\chi = 11{\cdot}1^\circ$, (*ii*) $\chi = 0{\cdot}5^\circ$).

3

BASIC KINETIC THEORY

3.1. The velocity distribution functions

A GAS flow would be completely described, in classical sense, by listings of the position, velocity, and internal state of every molecule at a particular instant. The number of molecules in a real gas is so large that such a description is unthinkable, and we must resort to a statistical description in terms of probability distributions. A variety of velocity distribution functions are employed in kinetic theory and possible confusion may be avoided by a general review of the relationships between them.

We will commence by defining the *single particle distribution function in velocity space*. Consider a sample of gas that is homogeneous in physical space and contains N identical molecules. A typical molecule has a velocity $\mathbf{c}$ with components u, v, and w in the directions of the Cartesian axes x, y, and z. Just as x, y, and z define a space called physical space, u, v, and w define *velocity space* as shown in Fig. 3.1. Each molecule can be represented in this space by a point defined by the appropriate velocity vector. The velocity distribution function $f(\mathbf{c})$ is then defined by

$$dN = Nf(\mathbf{c})\,du\,dv\,dw, \qquad (3.1)$$

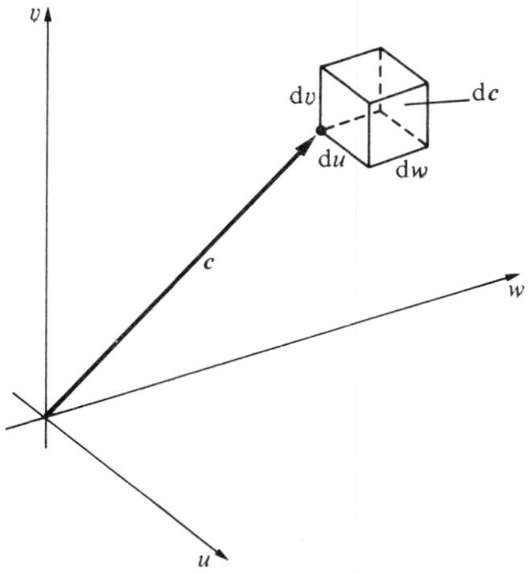

FIG. 3.1. Typical molecule and element in velocity space.

where dN is the number of molecules in the sample with velocity components u to $u+du$, v to $v+dv$ and w to $w+dw$. The product $du\,dv\,dw$ may be identified as a volume element in velocity space and is denoted by $d\mathbf{c}$. An alternative form of eqn (2.1) is, therefore,

$$dN = Nf(\mathbf{c})\,d\mathbf{c} \tag{3.1a}$$

and this need not be restricted to Cartesian coordinates. The functional statement is usually omitted and the function written simply as f. Also, since both dN and N refer to the molecules in the same volume of physical space, the number density may be used in place of the number. Therefore, the fraction of molecules within the velocity space element $d\mathbf{c}$ is

$$dn/n = f\,d\mathbf{c}. \tag{3.1b}$$

Since every molecule is represented by a point in velocity space,

$$\int_{-\infty}^{\infty}\int_{-\infty}^{\infty}\int_{-\infty}^{\infty} f\,du\,dv\,dw = \int_{-\infty}^{\infty} f\,d\mathbf{c} = \frac{N}{N} = 1. \tag{3.2}$$

The distribution function has therefore been normalized so that its integration over all velocity space yields unity. Note that f can never be negative and must either have finite bounds in velocity space or tend to zero as c tends to infinity.

The macroscopic quantities were defined in Chapter 1 in terms of averages of molecular velocities. The averages may be established as instantaneous, time, or ensemble averages over the molecules in an element of physical space. These molecules may be regarded as constituting a homogeneous gas sample and the single-particle distribution function in velocity space provides an appropriate description. In order to relate the macroscopic properties to this distribution function, we must determine the relationship between the function and the average value of any molecular quantity Q. This quantity is either a constant or a function of the molecular velocity. The mean value principle gives

$$\overline{Q} = \frac{1}{N}\int_N Q\,dN,$$

and, substituting for dN from eqn (3.1),

$$\overline{Q} = \frac{1}{N}\int_{-\infty}^{\infty} Qf(\mathbf{c})N\,d\mathbf{c}.$$

Omitting the functional statements, we have the general result

$$\overline{Q} = \int_{-\infty}^{\infty} Qf\,d\mathbf{c}. \tag{3.3}$$

This process is often referred to as establishing a *moment* of the distribution function and the macroscopic quantities are called moments of the distribution function. For example, the stream velocity which has been defined by eqn (1.15) as $\bar{\boldsymbol{c}}$ may be written

$$\bar{\boldsymbol{c}} = \int_{-\infty}^{\infty} \boldsymbol{c} f \, \mathrm{d}\boldsymbol{c}. \tag{3.4}$$

The macroscopic flow quantities are functions of position and time, and it is sometimes desirable to express the explicit dependence of the distribution function on the position vector in velocity space $\boldsymbol{r}$, and time t. Just as $\mathrm{d}\boldsymbol{c}$ has been used to denote a volume element in velocity space, a volume element in physical space may be denoted by $\mathrm{d}\boldsymbol{r}$. The product $\mathrm{d}\boldsymbol{c}\,\mathrm{d}\boldsymbol{r}$ then denotes a volume element in *phase space*, which is the multi-dimensional space formed by the combination of physical space and velocity space. The *single particle distribution function in phase space* $\mathscr{F}(\boldsymbol{c}, \boldsymbol{r}, t)$ is defined by

$$\mathrm{d}N = \mathscr{F}(\boldsymbol{c}, \boldsymbol{r}, t)\, \mathrm{d}\boldsymbol{c}\, \mathrm{d}\boldsymbol{r}, \tag{3.5}$$

with $\mathrm{d}N$ now representing the number of molecules in the phase-space element $\mathrm{d}\boldsymbol{c}\,\mathrm{d}\boldsymbol{r}$. In Cartesian coordinates, $\mathrm{d}\boldsymbol{c}\,\mathrm{d}\boldsymbol{r}$ becomes $\mathrm{d}u\,\mathrm{d}v\,\mathrm{d}w\,\mathrm{d}x\,\mathrm{d}y\,\mathrm{d}z$ and $\mathrm{d}N$ is the number of molecules with velocity components ranging from u to $u+\mathrm{d}u$, v to $v+\mathrm{d}v$, and w to $w+\mathrm{d}w$ and spatial coordinates ranging from x to $x+\mathrm{d}x$, y to $y+\mathrm{d}y$, and z to $z+\mathrm{d}z$. Note that $\mathscr{F}$ defines the number rather than the fraction of molecules in the phase space element. It has not been normalized and an integration over the whole phase space yields the total number of molecules in the system N, rather than unity.

If the distribution function in velocity space $f(\boldsymbol{c})$ is applied to the physical space element $\mathrm{d}\boldsymbol{r}$, the number of molecules N in eqn (3.1) represents the total number of molecules in $\mathrm{d}\boldsymbol{r}$, while $\mathrm{d}N$ represents the number of molecules in the phase-space element $\mathrm{d}\boldsymbol{c}\,\mathrm{d}\boldsymbol{r}$. We can then write

$$\mathrm{d}N = Nf(\boldsymbol{c})\,\mathrm{d}\boldsymbol{c} = \mathscr{F}(\boldsymbol{c}, \boldsymbol{r}, t)\,\mathrm{d}\boldsymbol{c}\,\mathrm{d}\boldsymbol{r}$$

and, since the number density in the phase space element is $N/\mathrm{d}\boldsymbol{r}$,

$$nf(\boldsymbol{c}) = \mathscr{F}(\boldsymbol{c}, \boldsymbol{r}, t).$$

Therefore, when $f(\boldsymbol{c})$ is used in a context in which it also depends on $\boldsymbol{r}$ and t, we have

$$nf \equiv \mathscr{F}. \tag{3.6}$$

We will take advantage of this identity to use f exclusively in the chapters that follow. Some authors (for example, Chapman and Cowling 1952 and Harris 1971) have preferred $\mathscr{F}$, which they have denoted by f. Therefore, in common with a number of other authors (for example, Kennard 1938 and Vincenti and Kruger 1965), our equations in nf will be equivalent to their equations in f.

The most specific distribution function is that for all N molecules in the system. At any instant, a complete system of monatomic molecules can be represented by a point in a $6N$-dimensional phase space. If we consider a large number or ensemble of such systems, the probability of finding a system in the volume element $\mathrm{d}\boldsymbol{c}_1\,\mathrm{d}\boldsymbol{c}_2 \ldots \mathrm{d}\boldsymbol{c}_N\,\mathrm{d}\boldsymbol{r}_1\,\mathrm{d}\boldsymbol{r}_2 \ldots \mathrm{d}\boldsymbol{r}_N$ about the phase space point $\boldsymbol{c}_1, \boldsymbol{r}_1, \boldsymbol{c}_2, \boldsymbol{r}_2, \ldots, \boldsymbol{c}_N, \boldsymbol{r}_N$ is

$$F^{(N)}(\boldsymbol{c}_1, \boldsymbol{r}_1, \boldsymbol{c}_2, \boldsymbol{r}_2, \ldots, \boldsymbol{c}_N, \boldsymbol{r}_N, t)\,\mathrm{d}\boldsymbol{c}_1\,\mathrm{d}\boldsymbol{c}_2 \ldots \mathrm{d}\boldsymbol{c}_N\,\mathrm{d}\boldsymbol{r}_1\,\mathrm{d}\boldsymbol{r}_2 \ldots \mathrm{d}\boldsymbol{r}_N, \qquad (3.7)$$

thus defining the *N particle distribution function* $F^{(N)}$. The subscript denotes the number of the molecule. A *reduced distribution function* $F^{(R)}$ for R of the N molecules is defined by

$$F^{(R)}(\boldsymbol{c}_1, \boldsymbol{r}_1, \boldsymbol{c}_2, \boldsymbol{r}_2, \ldots, \boldsymbol{c}_R, \boldsymbol{r}_R, t) = \int_{-\infty}^{\infty} \int_{-\infty}^{\infty} \times$$

$$\times\, F^{(N)}\,\mathrm{d}\boldsymbol{c}_{R+1} \ldots \mathrm{d}\boldsymbol{c}_N\,\mathrm{d}\boldsymbol{r}_{R+1} \ldots \mathrm{d}\boldsymbol{r}_N. \qquad (3.8)$$

In particular, the *single-particle distribution function* $F^{(1)}(\boldsymbol{c}_1, \boldsymbol{r}_1, t)$ is obtained by setting $R = 1$. The probability of finding molecule 1 in the phase space element $\mathrm{d}\boldsymbol{c}_1\,\mathrm{d}\boldsymbol{r}_1$ at time t is $F^{(1)}(\boldsymbol{c}_1, \boldsymbol{r}_1, t)$, irrespective of the positions of the other $N-1$ molecules. Since the molecules are indistinguishable, the number of molecules in the phase space element at time t is $NF^{(1)}$. We therefore have

$$NF^{(1)} \equiv \mathscr{F} \qquad (3.9)$$

and $F^{(1)}$ can be regarded as the normalized version of $\mathscr{F}$.

The *two-particle distribution function* $F^{(2)}(\boldsymbol{c}_1, \boldsymbol{r}_1, \boldsymbol{c}_2, \boldsymbol{r}_2, t)$ is of particular importance when considering binary collisions. The definition of a dilute gas requires that only a very small fraction of the space occupied by the gas actually contains a molecule. Therefore, in such a gas, it is generally assumed that the probability of finding a pair of molecules in a particular two-particle configuration is simply the product of the probabilities of finding the individual molecules in the two corresponding one-particle configurations. This requires

$$F^{(2)}(\boldsymbol{c}_1, \boldsymbol{r}_1, \boldsymbol{c}_2, \boldsymbol{r}_2, t) = F^{(1)}(\boldsymbol{c}_1, \boldsymbol{r}_1, t)F^{(1)}(\boldsymbol{c}_2, \boldsymbol{r}_2, t) \qquad (3.10)$$

and expresses the principle of *molecular chaos*. While the higher-order distribution functions are required for the study of dense gases, the single-particle distribution function provides an adequate description of dilute gases.

If the molecules are diatomic or polyatomic, the dimensions of phase space are increased by the number of internal degrees of freedom. Also, if the molecules are not spherically symmetric, their orientations must be specified. In general, the dimensions of phase space are equal to the least number of scalar variables required to specify the position, orientation, and internal state of a molecule. The definition of an extended distribution function to include these additional variables is quite straightforward as long as they are, or can be assumed to be, continuously distributed. This is the case as long as the classical description of the gas is valid. With the exception of hydrogen and helium at very low temperature, the translational motion can be described classically. The classical description is also valid for the rotational motion of gases such as oxygen and nitrogen at temperatures of the order of 300 K, but it is not valid for the vibrational modes in these gases. When discrete quantum states must be considered, a separate distribution function must be defined for each state. Separate distribution functions are also required for each species in a gas mixture. It is hardly surprising that most presentations of kinetic theory deal almost exclusively with simple gases consisting of spherically symmetric monatomic molecules.

3.2. The Boltzmann equation

We have seen that the velocity distribution functions provide a statistical description of a gas on the molecular level. The next step is establish the relationships between the distribution functions and the variables on which they depend. Ideally, the resulting equations would allow analytical solutions of problems in molecular gas dynamics.

The basic statistical mechanics equation for a gas is the Liouville equation which expresses the conservation of the N particle distribution function in $6N$-dimensional phase space. This equation is not directly useful, since the description of a real gas flow in terms of $F^{(N)}$ is completely out of the question. However, just as a hierarchy of reduced distribution functions $F^{(R)}$ were defined by eqn (3.8), a hierarchy of equations called the BBGKY equations may be obtained through the repeated integration of the Liouville equation. The final equation in the hierarchy is for the single particle distribution function $F^{(1)}$ and is the only one to hold out some hope of solution for practical gas flows. This equation also involves the two-particle distribution function $F^{(2)}$, but becomes a closed equation for $F^{(1)}$ when molecular chaos (eqn 3.10) is assumed. Then, through eqn (3.9), this becomes an equation for the single particle distribution function in phase space, and is equivalent to the equation that was originally formulated by Boltzmann (1872). The mathematical limits that define the validity of the Boltzmann equation are most precisely established through the above-mentioned derivation from the Liouville equation (see, for example: Grad (1958); Cercignani (1969); or Harris (1971)). On the other hand, the physical significance of each of the terms in the equation is more readily appreciated if the single particle distribution function is used throughout a derivation from first principles.

The latter procedure will be followed here and, for clarity and simplicity, the derivation will be restricted to a simple gas.

At a particular instant, the number of molecules in the phase space element $\mathrm{d}\boldsymbol{c}\,\mathrm{d}\boldsymbol{r}$ is given by eqn (3.5) as $\mathscr{F}\,\mathrm{d}\boldsymbol{c}\,\mathrm{d}\boldsymbol{r}$. The identity (3.6) enables this to be written $nf\,\mathrm{d}\boldsymbol{c}\,\mathrm{d}\boldsymbol{r}$. If the location and shape of the element do not vary with time, the rate of change of the number of molecules in the element is

$$\frac{\partial}{\partial t}(nf)\,\mathrm{d}\boldsymbol{c}\,\mathrm{d}\boldsymbol{r}. \tag{3.11}$$

The processes which contribute to the change in the number of molecules within $\mathrm{d}\boldsymbol{c}\,\mathrm{d}\boldsymbol{r}$ with time are illustrated in Fig. 3.2. They are:

(*i*) The convection of molecules across the surface of $\mathrm{d}\boldsymbol{r}$ by the molecular velocity $\boldsymbol{c}$. The representation of the phase space element as separate volume elements in physical and velocity space emphasizes the fact that $\boldsymbol{c}$ and $\boldsymbol{r}$ are treated as independent variables. $\boldsymbol{c}$ is regarded as constant within $\mathrm{d}\boldsymbol{r}$, and $\mathrm{d}\boldsymbol{c}$ is regarded as being located at the point defined by $\boldsymbol{r}$.

(*ii*) The 'convection' of molecules across the surface of $\mathrm{d}\boldsymbol{c}$ as a result of the external force per unit mass $\boldsymbol{F}$. The effect of the acceleration $\boldsymbol{F}$ on the molecules in $\mathrm{d}\boldsymbol{c}$ is analogous to the effect of the velocity $\boldsymbol{c}$ on the molecules in $\mathrm{d}\boldsymbol{r}$.

(*iii*) The scattering of molecules into and out of $\mathrm{d}\boldsymbol{c}\,\mathrm{d}\boldsymbol{r}$ as a result of intermolecular collisions. The gas is assumed to be dilute, as defined by eqn (1.5) and discussed in § 1.2. One consequence of this assumption is that a collision may be assumed to be an instantaneous event at a fixed location in physical space. This means that a molecule jumps from one point to another in velocity space, but remains at the same point in physical space and time.

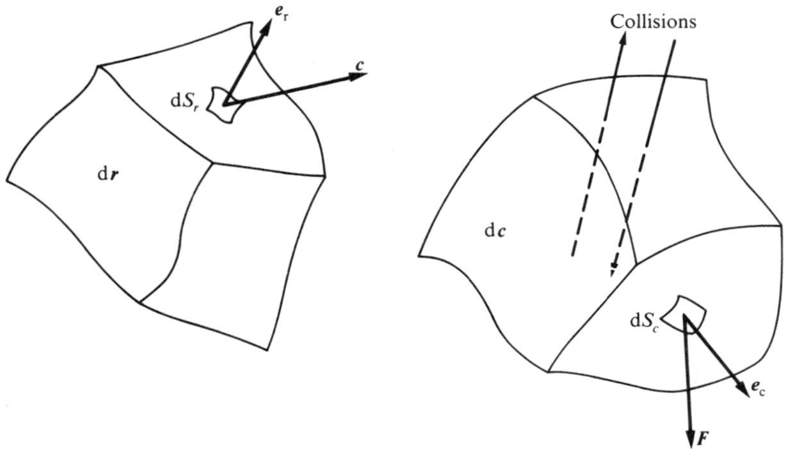

FIG. 3.2. Flux of molecules to and from a phase space element $\mathrm{d}\boldsymbol{c}\,\mathrm{d}\boldsymbol{r}$.

Therefore, in Fig. 3.2, collisions are represented as affecting only the element $\mathrm{d}\boldsymbol{c}$. A second major consequence of the dilute gas assumption is that all collisions may be assumed to be binary collisions.

First consider process (*i*) which is a convective effect across the surface of $\mathrm{d}\boldsymbol{r}$. The number of molecules in the phase space element is $nf\,\mathrm{d}\boldsymbol{c}\,\mathrm{d}\boldsymbol{r}$, so the number density of class $\boldsymbol{c}$ molecules within $\mathrm{d}\boldsymbol{r}$ is $nf\,\mathrm{d}\boldsymbol{c}$. Eqns (1.11) and (3.3) then enable the net inflow of molecules of this class across the surface of $\mathrm{d}\boldsymbol{r}$ to be written as

$$-\int_{S_r} nf\boldsymbol{c}\,.\,\boldsymbol{e}_r\,\mathrm{d}S_r\,\mathrm{d}\boldsymbol{c}.$$

Here, S_r is the total area of the surface of $\mathrm{d}\boldsymbol{r}$, $\mathrm{d}S_r$ is an element of this surface, and $\boldsymbol{e}_r$ is the unit normal vector of this element. Gauss' theorem enables the surface integral over S_r to be converted to a volume integral over $\mathrm{d}\boldsymbol{r}$. The expression then becomes

$$-\int_{d\boldsymbol{r}} \nabla\,.\,(nf\boldsymbol{c})\,\mathrm{d}(\mathrm{d}\boldsymbol{r})\,\mathrm{d}\boldsymbol{c}$$

or, since nf and $\boldsymbol{c}$ are constants within $\mathrm{d}\boldsymbol{r}$,

$$-\nabla\,.\,(nf\boldsymbol{c})\,\mathrm{d}\boldsymbol{r}\,\mathrm{d}\boldsymbol{c}.$$

Also, since we are considering only molecules of class $\boldsymbol{c}$, the velocity $\boldsymbol{c}$ may be taken outside the divergence in physical space. Therefore, the inflow of molecules of class $\boldsymbol{c}$ across the surface of $\mathrm{d}\boldsymbol{r}$ due to the velocity $\boldsymbol{c}$ is

$$-\boldsymbol{c}\,.\,\frac{\partial(nf)}{\partial\boldsymbol{r}}\,\mathrm{d}\boldsymbol{c}\,\mathrm{d}\boldsymbol{r}. \tag{3.12}$$

We may take advantage of the analogy between process (*ii*) in velocity space and process (*i*) in physical space to write the inflow of molecules across the surface of $\mathrm{d}\boldsymbol{c}$, due to the external force per unit mass $\boldsymbol{F}$, as

$$-\boldsymbol{F}\,.\,\frac{\partial(nf)}{\partial\boldsymbol{c}}\,\mathrm{d}\boldsymbol{c}\,\mathrm{d}\boldsymbol{r}. \tag{3.13}$$

The total number of molecules scattered out of the element $\mathrm{d}\boldsymbol{c}\,\mathrm{d}\boldsymbol{r}$ as a result of collisions is readily obtained through an analysis similar to that leading to eqns (1.6) and (1.7) for the collision frequency and total number of collisions in a gas. However, in order to obtain a meaningful expression for the molecules scattered into the element, we must consider both the pre-collision and post-collision velocities of the molecules participating in the collisions. In particular, we are concerned with the collision of a molecule of class $\boldsymbol{c}$ with one of class $\boldsymbol{c}_1$ such that their post-collision velocities are

$\boldsymbol{c}^*$ and $\boldsymbol{c}_1^*$, respectively. This is called a class $\boldsymbol{c}, \boldsymbol{c}_1 \to \boldsymbol{c}^*, \boldsymbol{c}_1^*$ collision, and we will now calculate the rate of scattering of molecules of class $\boldsymbol{c}$ out of $\mathrm{d}\boldsymbol{c}\,\mathrm{d}\boldsymbol{r}$ as a result of collisions of this class. A molecule of class $\boldsymbol{c}$ may be chosen as a test molecule moving with speed c_r among stationary field molecules of class $\boldsymbol{c}_1$. The volume swept out in physical space by the cross-section for this class of collision is $c_r\sigma\mathrm{d}\Omega$ and the number of class $\boldsymbol{c}_1$ molecules per unit volume in physical space is $nf_1\mathrm{d}\boldsymbol{c}_1$. The number of collisions of this class suffered by the test molecule per unit time is, therefore,

$$nf_1c_r\sigma\,\mathrm{d}\Omega\,\mathrm{d}\boldsymbol{c}_1.$$

Since the number of class $\boldsymbol{c}$ molecules in the phase space element is $nf\,\mathrm{d}\boldsymbol{c}\,\mathrm{d}\boldsymbol{r}$, the number of class $\boldsymbol{c}, \boldsymbol{c}_1 \to \boldsymbol{c}^*, \boldsymbol{c}_1^*$ collisions per unit time in the element is

$$n^2 f\,f_1 c_r\sigma\,\mathrm{d}\Omega\,\mathrm{d}\boldsymbol{c}_1\,\mathrm{d}\boldsymbol{c}\,\mathrm{d}\boldsymbol{r}. \tag{3.14}$$

Just as f denotes the value of the velocity distribution function f at $\boldsymbol{c}$, f_1 denotes the value of f at $\boldsymbol{c}_1$. Similarly, f^* and f_1^* may be used to denote the value of f at $\boldsymbol{c}^*$ and $\boldsymbol{c}_1^*$, respectively. Note also that the expression of a binary collision probability in terms of the product of two single-particle distribution functions has implicitly invoked the principle of molecular chaos.

The existence of inverse collisions (Fig. 2.2) means that an analysis, exactly similar to that leading to eqn (3.14), may be made for the collisions of class $\boldsymbol{c}^*, \boldsymbol{c}_1^* \to \boldsymbol{c}, \boldsymbol{c}_1$ that scatter molecules into class $\boldsymbol{c}$. This yields

$$n^2 f^* f_1^* c_r^* (\sigma\,\mathrm{d}\Omega)^*\,\mathrm{d}\boldsymbol{c}_1^*\,\mathrm{d}\boldsymbol{c}^*\,\mathrm{d}\boldsymbol{r} \tag{3.15}$$

for the collision rate in the phase-space element $\mathrm{d}\boldsymbol{c}^*\,\mathrm{d}\boldsymbol{r}$. Eqn (2.8) shows that c_r^* is equal to c_r, while the symmetry between the direct and inverse collisions is such that there is a unit Jacobean for the transformation between the pre-collision and post-collision values of the product of the differential cross section and velocity space elements. That is,

$$|(\sigma\,\mathrm{d}\Omega)\,\mathrm{d}\boldsymbol{c}_1\,\mathrm{d}\boldsymbol{c}| = |(\sigma\,\mathrm{d}\Omega)^*\,\mathrm{d}\boldsymbol{c}_1^*\,\mathrm{d}\boldsymbol{c}^*| \tag{3.16}$$

and eqn (3.15) may be written

$$n^2 f^* f_1^* c_r\sigma\,\mathrm{d}\Omega\,\mathrm{d}\boldsymbol{c}_1\,\mathrm{d}\boldsymbol{c}\,\mathrm{d}\boldsymbol{r}. \tag{3.17}$$

The rate of increase of molecules of class $\boldsymbol{c}$ in the phase space element $\mathrm{d}\boldsymbol{c}\,\mathrm{d}\boldsymbol{r}$ as a result of the combined direct and inverse collisions of class $\boldsymbol{c}, \boldsymbol{c}_1 \rightleftarrows \boldsymbol{c}^*, \boldsymbol{c}_1^*$ is obtained by subtracting the loss rate (expression (3.14)) from the rate of gain (expression 3.17). This gives

$$n^2(f^* f_1^* - f f_1) c_r\sigma\,\mathrm{d}\Omega\,\mathrm{d}\boldsymbol{c}_1\,\mathrm{d}\boldsymbol{c}\,\mathrm{d}\boldsymbol{r}. \tag{3.18}$$

The total rate of increase of molecules of class $\boldsymbol{c}$ in the element as a result of collisions is given by the integration of this expression over the complete cross section for its collisions with class $\boldsymbol{c}_1$ molecules, followed by the integration of the class $\boldsymbol{c}_1$ over all velocity space. The required expression for process (*iii*) is, therefore,

$$\int_{-\infty}^{\infty}\int_{0}^{4\pi} n^2(f^*f_1^* - ff_1)c_r\sigma \, d\Omega \, d\boldsymbol{c}_1 \, d\boldsymbol{c} \, d\boldsymbol{r}, \tag{3.19}$$

The expression (3.11) for the total rate of increase of molecules of class $\boldsymbol{c}$ due to all three processes may be equated to the sum of expression (3.19) for process (*iii*) and expressions (3.12) and (3.13) for processes (*i*) and (*ii*), respectively. If the latter terms are transferred to the left hand side and the complete equation is divided by $d\boldsymbol{c}\,d\boldsymbol{r}$, we have the *Boltzmann equation* for a simple dilute gas. This is

$$\frac{\partial}{\partial t}(nf) + \boldsymbol{c}\,.\,\frac{\partial}{\partial \boldsymbol{r}}(nf) + \boldsymbol{F}\,.\,\frac{\partial}{\partial \boldsymbol{c}}(nf) = \int_{-\infty}^{\infty}\int_{0}^{4\pi} n^2(f^*f_1^* - ff_1)c_r\sigma \, d\Omega \, d\boldsymbol{c}_1. \tag{3.20}$$

In a gas mixture consisting of a total of s chemical species, a separate distribution function must be defined for each species. The Boltzmann equation then becomes a set of s simultaneous equations and, as in § (1.3), particular species may be represented by the subscripts p or q. The Boltzmann equation for species p of the mixture can therefore be written

$$\frac{\partial}{\partial t}(n_p f_p) + \boldsymbol{c}_p\,.\,\frac{\partial}{\partial \boldsymbol{r}}(n_p f_p) + \boldsymbol{F}\,.\,\frac{\partial}{\partial \boldsymbol{c}}(n_p f_p)$$

$$= \sum_{q=1}^{s}\int_{-\infty}^{\infty}\int_{0}^{4\pi} n_p n_q(f_p^* f_{1q}^* - f_p f_{1q})c_{rpq}\sigma_{pq} \, d\Omega \, d\boldsymbol{c}_{1q}. \tag{3.21}$$

It was noted in the previous section that the presence of internal degrees of freedom requires the definition of extended distribution functions. Also, the collision cross-sections of asymmetric molecules are a function of the molecular orientation and therefore change continuously with time as the molecules rotate between collisions. Moreover, inverse collisions do not exist for the classical models of polyatomic molecules. It is, however, possible to define collision cross-sections that are smoothed or averaged over the molecular rotations and vibrations. The Liouville theorem then leads to a unit Jacobean for the transformation corresponding to eqn (3.16), thus permitting a final formulation similar to eqn (3.20). Chapman and Cowling (1970) have called the resulting equation the *generalized Boltzmann equation.*

The term on the right hand side of the Boltzmann equation is called the *collision term.* Its integral form contrasts with the partial differential form of the terms expressing the space and time dependence of nf, and is responsible for much of the mathematical difficulty associated with the Boltzmann equation. On the other hand, nf is the only dependent variable in the equation.

This might be considered an advantage when comparing the Boltzmann equation with the Navier–Stokes equations of continuum gas dynamics, since these have the velocity components and two thermodynamic properties as dependent variables. However, the advantage is far outweighted by the addition of the velocity-space coordinates to the list of independent variables. Although the Boltzmann equation was first put forward in 1872, forty years elapsed before solutions were obtained for the comparatively simple cases of uniform gradients of temperature, velocity, and species concentration in a gas. These are the Chapman–Enskog solutions and are discussed in § 3.5. They are restricted to low Knudsen number cases in which the mean free path is small in comparison with the scale length of the gradients. Serious attempts to solve the Boltzmann equation for more general flows at arbitrary Knudsen number date from about 1950; the various approaches are discussed in Chapter 6.

3.3. The moment and conservation equations

The quantity Q is associated with a single molecule and is either a constant or a function of the molecular velocity. We have previously seen that the average value of this quantity may be obtained through the multiplication of the velocity distribution function f by Q, followed by the integration of the product over all velocity space. These averages are referred to as moments of the distribution function. Similarly, a *moment of the Boltzmann equation* may be obtained by multiplying it by the quantity Q and then integrating the resulting equation over all velocity space. Since the moments of the distribution function include all the macroscopic quantities for a monatomic gas, the moment equations may be expected to include the monatomic gas version of the conservation equations of gas dynamics.

The multiplication of the Boltzmann equation (2.23) by Q yields

$$Q\frac{\partial}{\partial t}(nf)+Q\boldsymbol{c}\,.\,\frac{\partial}{\partial \boldsymbol{r}}(nf)+Q\boldsymbol{F}\,.\,\frac{\partial}{\partial \boldsymbol{c}}(nf)$$

$$= Q\int_{-\infty}^{\infty}\int_{0}^{4\pi} n^2(f^*f_1^*-ff_1)c_r\sigma\,\mathrm{d}\Omega\,\mathrm{d}\boldsymbol{c}_1\,. \tag{3.22}$$

Both f and Q refer to molecules of class $\boldsymbol{c}$ and the moment equation is obtained by integrating over all classes of molecule. Since Q is either a constant or a function of $\boldsymbol{c}$ only, it may be taken within the derivative in the first term. The required integral of this term is, therefore,

$$\int_{-\infty}^{\infty}\frac{\partial}{\partial t}(nQf)\mathrm{d}\boldsymbol{c}$$

or, using eqn (3.3),

$$\frac{\partial}{\partial t}(n\overline{Q}). \tag{3.23}$$

Both $\boldsymbol{c}$ and Q may be taken inside the derivative in the second term of eqn (3.22), the integral becoming

$$\int_{-\infty}^{\infty} \nabla \cdot (n\boldsymbol{c}Qf)\,\mathrm{d}\boldsymbol{c}$$

or

$$\nabla \cdot (n\overline{\boldsymbol{c}Q}). \tag{3.24}$$

The reason why $\boldsymbol{c}$ may be taken inside the derivative in eqn (3.22) is that the dependent variable in the Boltzmann equation is the value of the distribution function for this particular class of molecule. Therefore, while f is a function of $\boldsymbol{r}$ and t, $\boldsymbol{c}$ and Q may be regarded as being independent of $\boldsymbol{r}$ and t in eqn (3.22). However, the averages of $\boldsymbol{c}$ and Q are established through the distribution function f and must be treated as functions of $\boldsymbol{r}$ and t. The integral of the third term in eqn (3.22) is

$$\int_{-\infty}^{\infty} Q\boldsymbol{F} \cdot \frac{\partial}{\partial \boldsymbol{c}}(nf)\,\mathrm{d}\boldsymbol{c},$$

which may be written

$$\int_{-\infty}^{\infty} \boldsymbol{F} \cdot \frac{\partial}{\partial \boldsymbol{c}}(nQf)\,\mathrm{d}\boldsymbol{c} - \int_{-\infty}^{\infty} \boldsymbol{F} \cdot \frac{\partial Q}{\partial \boldsymbol{c}} nf\,\mathrm{d}\boldsymbol{c}.$$

It is assumed that $\boldsymbol{F}$ is independent of $\boldsymbol{c}$ and, since $f = 0$ or $f \to 0$ as $c \to \infty$, the first integral vanishes and the second becomes

$$-n\boldsymbol{F} \cdot \overline{\frac{\partial Q}{\partial \boldsymbol{c}}}. \tag{3.25}$$

The integral of the term on the right hand side of eqn (3.22) is called the *collision integral* and is denoted by $\Delta[Q]$, i.e.

$$\Delta[Q] = \int_{-\infty}^{\infty}\int_{-\infty}^{\infty}\int_{0}^{4\pi} n^2 Q(f^* f_1^* - f f_1) c_r \sigma\,\mathrm{d}\Omega\,\mathrm{d}\boldsymbol{c}_1\,\mathrm{d}\boldsymbol{c}. \tag{3.26}$$

Eqns (3.23) to (3.26) may now be brought together to write the moment equation for Q as

$$\frac{\partial}{\partial t}(n\overline{Q}) + \nabla \cdot (n\overline{\boldsymbol{c}Q}) - n\boldsymbol{F} \cdot \overline{\frac{\partial Q}{\partial \boldsymbol{c}}} = \Delta[Q]. \tag{3.27}$$

This equation may also be called the *transfer equation* or the *equation of change.*

Two symmetries are associated with the collision term and they lead to several alternative forms of $\Delta[Q]$. These clarify the physical meaning of the collision term and will be required for future applications. The first symmetry is between the collision partners and means that $\Delta[Q]$ is unchanged if $\boldsymbol{c}$ and $\boldsymbol{c}_1$ are interchanged and Q, which represents the value of Q at $\boldsymbol{c}$, is interchanged with the value Q_1 at $\boldsymbol{c}_1$. Similarly $\boldsymbol{c}^*$ and $\boldsymbol{Q}^*$ may be interchanged with $\boldsymbol{c}_1^*$ and $\boldsymbol{Q}_1^*$. The second symmetry is based on the existence of inverse collisions and is between the pre-collision and post-collision velocities. It enables Q and Q^* or Q_1 and Q_1^* to be interchanged, as long as $\boldsymbol{c}_1$, $\boldsymbol{c}$, f_1, and f are interchanged with $\boldsymbol{c}_1^*$, $\boldsymbol{c}^*$, f_1^*, and f^*. Advantage is taken of eqn (3.16) to avoid replacing $\mathrm{d}\boldsymbol{c}\,\mathrm{d}\boldsymbol{c}_1$ by $\mathrm{d}\boldsymbol{c}^*\,\mathrm{d}\boldsymbol{c}_1^*$. Application of the first symmetry to eqn (3.26) yields

$$\Delta[Q] = \int_{-\infty}^{\infty}\int_{-\infty}^{\infty}\int_{0}^{4\pi} n^2 Q_1(f_1^* f^* - f_1 f) c_r \sigma \,\mathrm{d}\Omega\,\mathrm{d}\boldsymbol{c}\,\mathrm{d}\boldsymbol{c}_1. \tag{3.26a}$$

The second symmetry may then be applied to this equation to give

$$\Delta[Q] = \int_{-\infty}^{\infty}\int_{-\infty}^{\infty}\int_{0}^{4\pi} n^2 Q_1^*(f_1 f - f_1^* f^*) c_r \sigma \,\mathrm{d}\Omega\,\mathrm{d}\boldsymbol{c}\,\mathrm{d}\boldsymbol{c}_1, \tag{3.26b}$$

and a second application of the first symmetry gives

$$\Delta[Q] = \int_{-\infty}^{\infty}\int_{-\infty}^{\infty}\int_{0}^{4\pi} n^2 Q^*(f f_1 - f^* f_1^*) c_r \sigma \,\mathrm{d}\Omega\,\mathrm{d}\boldsymbol{c}\,\mathrm{d}\boldsymbol{c}_1. \tag{3.26c}$$

Eqns (3.26), (3.26*a*), (3.26*b*), and (3.26*c*) may then be summed and the resultant equation divided by four, to give

$$\Delta[Q] = \tfrac{1}{4}\int_{-\infty}^{\infty}\int_{-\infty}^{\infty}\int_{0}^{4\pi} n^2 (Q + Q_1 - Q^* - Q_1^*)(f^* f_1^* - f f_1) c_r \sigma \,\mathrm{d}\Omega\,\mathrm{d}\boldsymbol{c}_1\,\mathrm{d}\boldsymbol{c}. \tag{3.26d}$$

Also, eqn (3.26) may be written

$$\Delta[Q] = \int_{-\infty}^{\infty}\int_{-\infty}^{\infty}\int_{0}^{4\pi} n^2 Q f^* f_1^* c_r \sigma \,\mathrm{d}\Omega\,\mathrm{d}\boldsymbol{c}_1\,\mathrm{d}\boldsymbol{c} - \int_{-\infty}^{\infty}\int_{-\infty}^{\infty}\int_{0}^{4\pi} \times$$
$$\times\, n^2 Q f f_1 c_r \sigma \,\mathrm{d}\Omega\,\mathrm{d}\boldsymbol{c}_1\,\mathrm{d}\boldsymbol{c},$$

and application of the second symmetry to the first term on the right hand side converts $Qf^*f_1^*$ to Q^*ff_1. Therefore,

$$\Delta[Q] = \int_{-\infty}^{\infty}\int_{-\infty}^{\infty}\int_{0}^{4\pi} n^2 (Q^* - Q) f f_1 c_r \sigma \,\mathrm{d}\Omega\,\mathrm{d}\boldsymbol{c}_1\,\mathrm{d}\boldsymbol{c}. \tag{3.26e}$$

A similar transformation of eqn (3.26) yields

$$\Delta[Q] = \int_{-\infty}^{\infty}\int_{-\infty}^{\infty}\int_{0}^{4\pi} n^2(Q_1^* - Q_1) f_1 f c_r \sigma \, d\Omega \, d\boldsymbol{c}_1 \, d\boldsymbol{c}. \qquad (3.26f)$$

Finally, eqns (3.26*e*) and (3.26*f*) may be summed and the resulting equation halved, to give

$$\Delta[Q] = \tfrac{1}{2}\int_{-\infty}^{\infty}\int_{-\infty}^{\infty}\int_{0}^{4\pi} n^2(Q^* + Q_1^* - Q - Q_1) f_1 f c_r \sigma \, d\Omega \, d\boldsymbol{c} \, d\boldsymbol{c}_1. \; (3.26g)$$

The physical meaning of $\Delta[Q]$ is most readily apparent from eqn (3.26g), since $(Q^* + Q_1^* - Q - Q_1)$ represents the change in the quantity Q as the result of a collision of class $\boldsymbol{c}, \boldsymbol{c}_1 \rightarrow \boldsymbol{c}^*\boldsymbol{c}_1^*$. This change is summed over all classes of collision and halved to allow for the double counting of collisions in the integration. That the integration is equivalent to summing over all collisions is readily seen if we note that

$$\tfrac{1}{2}\int_{-\infty}^{\infty}\int_{-\infty}^{\infty}\int_{0}^{4\pi} n^2 f_1 f c_r \sigma \, d\Omega \, d\boldsymbol{c} \, d\boldsymbol{c}_1 = \tfrac{1}{2}\int_{-\infty}^{\infty}\int_{-\infty}^{\infty} \sigma_T c_r n^2 f_1 f \, d\boldsymbol{c}_1 \, d\boldsymbol{c}$$
$$= \tfrac{1}{2} n^2 \overline{\sigma_T c_r},$$

which is in agreement with eqn (1.7) for the total number of collisions per unit time per unit volume of gas. The moment or transfer equation with $\Delta[Q]$ in the form of eqn (3.26g) can therefore be derived independently of the Boltzmann equation and was, in fact, first derived by Maxwell (1867). The form of $\Delta[Q]$ in eqn (3.26*d*) is more closely allied to the Boltzmann formulation and states simply that the change in Q as a result of the inverse collisions is exactly equal to the change in Q as a result of the direct collisions.

If Q is either the mass m, momentum $m\boldsymbol{c}$, or energy $\tfrac{1}{2}mc^2$ of a molecule, the conservation of these quantities in collisions requires that $Q + Q_1 - Q^* - Q_1^* = 0$. Eqns (3.26*d*) and (3.26g) then show that the collision integral $\Delta[Q]$ is zero, as would be expected from the physical meaning of the integral. The quantities m, $m\boldsymbol{c}$, and $\tfrac{1}{2}mc^2$ are called *collisional invariants*, while any quantity that satisfies the condition $Q + Q_1 - Q^* - Q_1^* = 0$ is called a *summational invariant*. It can be shown (e.g. Harris (1971) § 4.2) that the collisional invariants, or linear combinations of them, are the only summational invariants. Therefore, if Q is a summational invariant, the collision integral $\Delta[Q]$ is zero and Q can be written

$$Q = A\tfrac{1}{2}mc^2 + \boldsymbol{B} \,.\, m\boldsymbol{c} + C, \qquad (3.28)$$

where A, $\boldsymbol{B}$, and C are constants.

The collisional integral is zero in the three equations for the collisional invariants and the averages on the left hand side of the equations can be expressed in terms of the macroscopic gas quantities. The three equations are the *conservation equations* of gas dynamics. First, the equation for the *conservation of mass* is obtained by setting $Q = m$ in eqn (3.27), to give

$$\frac{\partial}{\partial t}(nm)+\nabla\cdot(nm\bar{c}) = 0. \tag{3.29}$$

or, using eqns (1.9) and (1.13),

$$\frac{\partial \rho}{\partial t}+\nabla\cdot(\rho \boldsymbol{c}_0) = 0. \tag{3.30}$$

It is convenient to introduce the *substantial derivative*

$$\frac{D}{Dt} \equiv \frac{\partial}{\partial t}+\boldsymbol{c}_0\cdot\nabla \equiv \frac{\partial}{\partial t}+u\frac{\partial}{\partial x}+v\frac{\partial}{\partial y}+w\frac{\partial}{\partial z} \tag{3.31}$$

to denote differentiation following the motion of a fluid element. The equation for the conservation of mass, or continuity equation, then becomes

$$\frac{D\rho}{Dt} = -\rho\nabla\cdot\boldsymbol{c}_0. \tag{3.32}$$

Next, the equation for the *conservation of momentum* or *equation of motion*, is obtained by setting $Q = m\boldsymbol{c}$ in eqn (3.27). This is the vector equation

$$\frac{\partial}{\partial t}(\rho\bar{\boldsymbol{c}})+\nabla\cdot(\rho\overline{\boldsymbol{c}\boldsymbol{c}})-\rho\boldsymbol{F} = 0. \tag{3.33}$$

But eqns (1.15) and (1.16) show that

$$\overline{\boldsymbol{c}\boldsymbol{c}} = \overline{(\boldsymbol{c}'+\boldsymbol{c}_0)(\boldsymbol{c}'+\boldsymbol{c}_0)} = \overline{\boldsymbol{c}'\boldsymbol{c}'}+\boldsymbol{c}_0\boldsymbol{c}_0$$

and eqn (3.33) may be written

$$\rho\frac{\partial \boldsymbol{c}_0}{\partial t}+\boldsymbol{c}_0\frac{\partial \rho}{\partial t}+\boldsymbol{c}_0\nabla\cdot(\rho\boldsymbol{c}_0)+\rho(\boldsymbol{c}_0\cdot\nabla)\boldsymbol{c}_0+\nabla\cdot(\rho\overline{\boldsymbol{c}'\boldsymbol{c}'})-\rho\boldsymbol{F} = 0.$$

The continuity equation (3.30) enables the second and third terms to be removed, eqn (3.31) enables the first and fourth terms to be combined as a substantial derivative, and eqn (1.20) enables the fifth term to be written in terms of the pressure and the viscous stress tensor, i.e.

$$\rho\frac{D\boldsymbol{c}_0}{Dt} = -\nabla p+\nabla\cdot\boldsymbol{\tau}+\rho\boldsymbol{F}. \tag{3.34}$$

The vector and tensor quantities are clarified if we also write the momentum equation in the x direction in Cartesian coordinates. This is

$$\rho\frac{Du}{Dt} = -\frac{\partial p}{\partial x}+\frac{\partial \tau_{xx}}{\partial x}+\frac{\partial \tau_{xy}}{\partial y}+\frac{\partial \tau_{xz}}{\partial z}+\rho F_x.$$

Finally, Q may be set equal to $\frac{1}{2}mc^2$ in eqn (3.27) to give the equation for the *conservation of energy* as

$$\frac{\partial}{\partial t}(\tfrac{1}{2}\rho\overline{c^2})+\nabla\,.\,(\tfrac{1}{2}\rho\overline{\boldsymbol{c}c^2})-\rho\boldsymbol{c}_0\,.\,\boldsymbol{F}=0. \tag{3.35}$$

A similar process to that applied to eqn (3.33) (see Exercise 1.9 for the expansion of $\overline{\boldsymbol{c}c^2}$) finally yields

$$\rho\frac{De}{Dt}=-p\nabla\,.\,\boldsymbol{c}_0+\Phi-\nabla\,.\,\boldsymbol{q}. \tag{3.36}$$

The quantity Φ is called the dissipation function and may be written in Cartesian coordinates as

$$\Phi=\tau_{xx}\frac{\partial u}{\partial x}+\tau_{yy}\frac{\partial v}{\partial y}+\tau_{zz}\frac{\partial w}{\partial z}+\tau_{xy}\left(\frac{\partial u}{\partial y}+\frac{\partial v}{\partial x}\right)+$$
$$+\tau_{yz}\left(\frac{\partial v}{\partial z}+\frac{\partial w}{\partial y}\right)+\tau_{zx}\left(\frac{\partial w}{\partial x}+\frac{\partial u}{\partial z}\right). \tag{3.37}$$

Since the momentum equation (3.34) is a vector equation, the equations for the conservation of mass, momentum, and energy constitute five equations. In addition to the three velocity components and the three thermodynamic properties p, ρ, and T, they contain the viscous stress tensor $\boldsymbol{\tau}$ and heat flux vector $\boldsymbol{q}$ as dependent variables. They do not, therefore, form a determinate set. However, if $\boldsymbol{\tau}$ and $\boldsymbol{q}$ are both zero, the addition of the equation of state results in a determinate set. These are the *Euler equations* of inviscid flow.

The application of moment equations other than the conservation equations generally requires the evaluation of the collision integral $\Delta[Q]$. This process may be illustrated by the special case of $Q=u^2$ for Maxwell molecules. The substitution of this value of Q into eqn (3.26*e*), together with eqn (2.33) for the cross-section of Maxwell molecules, yields

$$\Delta[u^2]=\int_{-\infty}^{\infty}\int_{-\infty}^{\infty}\int_{0}^{2\pi}\int_{0}^{\infty}n^2(u^{*2}-u^2)ff_1W_0\left(\frac{2\kappa}{m}\right)^{\frac{1}{2}}\mathrm{d}W_0\,\mathrm{d}\varepsilon\,\mathrm{d}\boldsymbol{c}\,\mathrm{d}\boldsymbol{c}_1.$$

The advantage of Maxwell molecules is that the molecular velocities appear explicitly only in the term $u^{*2}-u^2$, which may be written as $(u^*-u)^2+ +2u(u^*-u)$. Then with u and u_1 identified with u_2 and u_1 in eqns (2.4) and (2.5), and using eqn (2.22),

$$u^*-u=\tfrac{1}{2}(u_r-u_r^*)=\tfrac{1}{2}\{(1-\cos\chi)u_r-\sin\chi\sin\varepsilon(v_r^2+w_r^2)^{\frac{1}{2}}\}.$$

Therefore,

$$u^{*2}-u^2 = \tfrac{1}{4}\{(1-\cos\chi)^2u_r^2-2(1-\cos\chi)u_r\sin\chi\sin\varepsilon(v_r^2+w_r^2)^{\frac{1}{2}}+\sin^2\chi\times$$
$$\times\sin^2\varepsilon(v_r^2+w_r^2)\}+u\{(1-\cos\chi)u_r-\sin\chi\sin\varepsilon(v_r^2+w_r^2)^{\frac{1}{2}}\}.$$

The integration over ε may be carried out first, to give,

$$\int_0^{2\pi}(u^{*2}-u^2)\,\mathrm{d}\varepsilon = \pi(u_r^2+2u\,u_r)(1-\cos\chi)-\tfrac{1}{4}\pi(3u_r^2-c_r^2)\sin^2\chi.$$

Eqn (2.3) shows that $u_r^2+2uu_r = u_1^2-u^2$ and, since u and u_1 are described by the same distribution, this term disappears in the integration over velocity space. The remaining term contains only relative velocities, which means that the result is independent of whether these are expressed as differences between velocities or peculiar velocities and, therefore, $\Delta[u^2] = \Delta[u'^2]$. Eqns (2.3) and (1.16) give $\underline{u_r^2} = u_1'-2u_1'u'+u'^2$ and the double integration over velocity space yields $2\overline{u'^2}$. A similar result holds for c_r^2, and the collision integral becomes

$$\Delta[u^2] = -\frac{3\pi}{2}\left(\frac{2\kappa}{m}\right)^{\frac{1}{2}}n^2(\overline{u'^2}-\tfrac{1}{3}\overline{c'^2})\int_0^\infty\sin^2\chi\,W_0\,\mathrm{d}W_0.$$

Finally, from eqns (1.19) and (1.20),

$$\Delta[u^2] = \frac{3\pi}{2}A_2(5)\left(\frac{2\kappa}{m}\right)^{\frac{1}{2}}\frac{n}{m}\tau_{xx}, \tag{3.38}$$

where

$$A_2(5) = \int_0^\infty\sin^2\chi\,W_0\,\mathrm{d}W_0$$

for Maxwell molecules. $A_2(5)$ is a pure number and its value may be obtained through eqn (2.32) as 0·436.

3.4. The *H*-theorem and equilibrium

Consider a spatially homogeneous volume of a simple dilute monatomic gas that is free of any external force. The Boltzmann equation (3.20) may be simplified for such a gas since the number density n is a constant, spatial derivatives $\partial/\partial\boldsymbol{r}$ are zero, and the external force $\boldsymbol{F}$ is zero. The equation becomes

$$\frac{\partial f}{\partial t} = n\int_{-\infty}^{\infty}\int_0^{4\pi}(f^*f_1^*-ff_1)c_r\sigma\,\mathrm{d}\Omega\,\mathrm{d}\boldsymbol{c}_1. \tag{3.39}$$

Over a small time interval, f changes to $f+\Delta f$ and the fractional change is

$\Delta f / f$ or $\Delta(\ln f)$. Boltzmann's H-function is the mean value of $\ln f$, i.e.

$$H = \overline{\ln f}$$

or, using eqn (3.3),

$$H = \int_{-\infty}^{\infty} f \ln f \, \mathrm{d}\boldsymbol{c}. \tag{3.40}$$

The quantity Q in the moment equation (3.27) may then be set equal to $\ln f$ and, with the collision integral $\Delta[Q]$ in the form given by eqn (3.26*d*), we have

$$\frac{\partial H}{\partial t} = \frac{n}{4} \int_{-\infty}^{\infty} \int_{-\infty}^{\infty} \int_{0}^{4\pi} (\ln f + \ln f_1 - \ln f^* - \ln f_1^*)(f^* f_1^* - f f_1) c_r \sigma \, \mathrm{d}\Omega \, \mathrm{d}\boldsymbol{c} \, \mathrm{d}\boldsymbol{c}_1$$

$$= \frac{n}{4} \int_{-\infty}^{\infty} \int_{-\infty}^{\infty} \int_{0}^{4\pi} \ln (f f_1 / f^* f_1)(f^* f_1^* - f f_1) c_r \sigma \, \mathrm{d}\Omega \, \mathrm{d}\boldsymbol{c} \, \mathrm{d}\boldsymbol{c}_1. \tag{3.41}$$

If $\ln (f f_1 / f^* f_1^*)$ is positive, then $(f^* f_1^* - f f_1)$ must be negative and vice versa. The integral on the right hand side of eqn (3.14) is, therefore, either negative or zero and H can never increase, i.e.

$$\frac{\partial H}{\partial t} < 0. \tag{3.42}$$

This result is known as Boltzmann's H-theorem.

The question which now arises is whether H decreases without limit to $-\infty$, or tends to a finite value and thereafter remains constant. Now, as $c \to \infty$, $f \to 0$ and $\ln f \to -\infty$, and it appears that the integral for H in eqn (3.40) may diverge. However, since the energy of the gas is finite, the integral

$$\int_{-\infty}^{\infty} f c^2 \, \mathrm{d}\boldsymbol{c}$$

must converge. Therefore, for H to diverge, the approach of $\ln f$ to $-\infty$ must be more rapid than that of $-c^2$ to $-\infty$. In this case, the approach of f to 0 would be more rapid than that of $\exp(-c^2)$ to 0 and, since $\exp(-x^2)$ $x^n \to 0$ as $x \to \infty$ for all the values of n, H must converge. Therefore, for any initial distribution of molecules in velocity space, the distribution will alter with time in such a way that H decreases monatonically to a finite lower bound. At subsequent times,

$$\frac{\partial H}{\partial t} = 0$$

and eqn (2.41) shows that this requires

$$f^*f_1^* - ff_1 = 0 \tag{3.43}$$

or, equivalently,

$$\ln f + \ln f_1 = \ln f^* + \ln f_1^*. \tag{3.44}$$

A comparison of eqns (3.44) and (3.39) shows that the stationary state for H is also a stationary state for f. We therefore have an *equilibrium state* in which the probable number of molecules in any element of velocity space remains constant with time.

Eqn (3.44) shows that, in the equilibrium state, $\ln f$ is a summational invariant. Therefore, from eqn (3.28), the necessary and sufficient condition for equilibrium is that

$$\ln f = A\tfrac{1}{2}mc^2 + \boldsymbol{B}\,.\,m\boldsymbol{c} + C. \tag{3.45}$$

This equation may be written in terms of the thermal velocity components as

$$\ln f = A\tfrac{1}{2}mc'^2 + m(A\boldsymbol{c}_0 + \boldsymbol{B})\,.\,\boldsymbol{c}' + A\tfrac{1}{2}mc_0^2 + \boldsymbol{B}\,.\,m\boldsymbol{c}_0 + C.$$

Since there can be no preferred direction in the equilibrium gas, the distribution must be isotropic. This requires that the coefficient of c' be zero, or

$$\boldsymbol{B} = -A\boldsymbol{c}_0.$$

Therefore,

$$\ln f = \tfrac{1}{2}Amc'^2 - \tfrac{1}{2}Amc_0^2 + C,$$

or,

$$f = \exp(\tfrac{1}{2}Amc'^2 - \tfrac{1}{2}Amc_0^2 + C).$$

Since f is bounded, the coefficient of c'^2 must be negative and, for convenience, we introduce a new constant by setting $\tfrac{1}{2}Am = -\beta^2$. We therefore have

$$f = \exp(C + \beta^2 c_0^2)\exp(-\beta^2 c'^2).$$

The constant C may be eliminated through the normalization condition expressed in eqn (3.2). This requires that

$$\int_{-\infty}^{\infty} f\,\mathrm{d}\boldsymbol{c} = \exp(C + \beta^2 c_0^2)\int_{-\infty}^{\infty}\exp(-\beta^2 c'^2)\,\mathrm{d}\boldsymbol{c}' = 1.$$

But,

$$\int_{-\infty}^{\infty}\exp(-\beta^2 c'^2)\,\mathrm{d}\boldsymbol{c}' = \int_{-\infty}^{\infty}\int_{-\infty}^{\infty}\int_{-\infty}^{\infty}\exp\{-\beta^2(u'^2+v'^2+w'^2)\}\,\mathrm{d}u'\,\mathrm{d}v'\,\mathrm{d}w'$$

$$= \int_{-\infty}^{\infty}\exp(-\beta^2 u'^2)\,\mathrm{d}u'\int_{-\infty}^{\infty}\exp(-\beta^2 v'^2)\,\mathrm{d}v'\int_{-\infty}^{\infty}\exp(-\beta^2 w'^2)\,\mathrm{d}w'.$$

The list of standard integrals in Appendix B shows that each component integral is $\surd(\pi)/\beta$, so that

$$\exp(C+\beta^2 c_0^2) = \beta^3/\pi^{\frac{3}{2}}.$$

Therefore, the final result for the *equilibrium* or *Maxwellian distribution function* f_0 is,

$$f_0 = (\beta^3/\pi^{\frac{3}{2}}) \exp(-\beta^2 c'^2). \tag{3.46}$$

The constant β may be related to the temperature of the gas. Eqns (1.23), (3.4), and (3.46) show that

$$\tfrac{3}{2}RT = \tfrac{1}{2}\overline{c'^2} = \frac{\beta^3}{2\pi^{\frac{3}{2}}} \int_{-\infty}^{\infty} c'^2 \exp(-\beta^2 c'^2)\, \mathrm{d}\mathbf{c}$$

or again using the standard integrals of Appendix B,

$$\beta^2 = (2RT)^{-1} = m/(2kT). \tag{3.47}$$

Since we are dealing with an equilibrium gas, the translational temperature T_{tr} calculated through eqn (1.23) is equal to the thermodynamic temperature T and the subscript has been dropped. Substitution of eqn (3.47) into (3.46) gives

$$f_0 = \left(\frac{m}{2\pi kT}\right)^{\frac{3}{2}} \exp\{-mc'^2/(2kT)\} \tag{3.46a}$$

as an alternative definition of f_0.

In the case of a gas mixture in equilibrium, it can be shown (Chapman and Cowling 1952, p. 84) that the distribution function of species p follows directly as

$$f_{0,p} = \left(\frac{m_p}{2\pi kT}\right)^{\frac{3}{2}} \{-m_p c_p'^2/(2kT)\}. \tag{3.48}$$

The Maxwellian distribution applies also to the translational velocities of diatomic or polyatomic molecules. This result has been proven for the special case of rough-spherical molecules (Chapman and Cowling 1952; § 11.4) using an extension of the method that has been applied here to monatomic molecules. The collision mechanics of more realistic models are sufficiently complex to rule out this approach. However, if it is assumed that, for each collision, a collision with the velocity components reversed is equally likely, the generalized Boltzmann equation leads to a derivation applicable to all molecular models (Chapman and Cowling 1970; § 11.3). Moreover, for equilibrium situations, the methods of statistical mechanics are available and provide proofs that are more general than those from kinetic theory.

We will usually be dealing with gas flows that are neither spatially homogeneous nor macroscopically steady, and it is necessary to define the circumstances under which the Maxwellian distribution may be applied to such flows. If the distribution of velocities at each location in a gas flow is in accordance with the Maxwellian distribution based on the local temperature and density, the flow is said to be in *local thermodynamic equilibrium.* This concept of local thermodynamics equilibrium is, in fact, identical with the more familiar concept of *thermodynamic reversibility.* As a general rule we will find that a flow with zero viscous stresses, heat fluxes, and diffusion velocities may be treated as reversible as long as

$$(1/\nu)D(\ln Q)/Dt \ll 1. \tag{3.49}$$

Here, the quantity Q represents any macroscopic quantity. The condition (3.49) states that the time scales for the variations of the macroscopic quantities in a frame of reference moving with a fluid element must be extremely large in comparison with the mean collision time. However, it should be noted that a process involving finite heat transfer, viscous stress, or diffusion is irreversible, even though $(1/\nu)D(\ln Q)/Dt$ may be zero for all Q. The most obvious example is provided by a stationary gas with zero pressure gradients, but finite temperature and density gradients. A gas flow in local thermodynamic equilibrium is described by the Euler equations.

3.5. The Chapman–Enskog method

The Chapman–Enskog method provides a solution of the Boltzmann equation for a restricted set of problems in which the distribution function f is perturbed by a small amount from the equilibrium Maxwellian form. It is assumed that the distribution function may be expressed in the form of the power series.

$$f = f^{(0)} + \varepsilon_0 f^{(1)} + \varepsilon_0^2 f^{(2)} + \dots, \tag{3.50}$$

where ε_0 is a parameter which may be regarded as a measure of either the mean collision time or the Knudsen number. The first term $f^{(0)}$ is the Maxwellian distribution f_0 for an equilibrium gas and an alternative form of the expression is

$$f = f_0(1 + \Phi_1 + \Phi_2 + \dots). \tag{3.51}$$

The equilibrium distribution function constitutes the known first-order solution of this equation, and the second-order solution requires the determination of the parameter Φ_1.

Solutions of the Boltzmann equation for $f = f_0(1+\Phi_1)$ were obtained independently by Chapman and Enskog, and these form the major subject matter of the classical work by Chapman and Cowling (1952). For a simple gas, Φ_1 depends only on the density, stream velocity, and temperature of the gas, so that the resulting solution constitutes a *normal solution* of the Boltzmann equation. Since f_0 satisfies the equations

$$\int_{-\infty}^{\infty} f\,\mathrm{d}\boldsymbol{c} = 1,$$

$$\int_{-\infty}^{\infty} \boldsymbol{c}f\,\mathrm{d}\boldsymbol{c} = \boldsymbol{c}_0,$$

and

$$\int_{-\infty}^{\infty} c'^2 f\,\mathrm{d}\boldsymbol{c} = 3RT,$$

Φ_1 must be such that

$$\int_{-\infty}^{\infty} \Phi_1 f_0\,\mathrm{d}\boldsymbol{c} = 0,$$

$$\int_{-\infty}^{\infty} \boldsymbol{c}\Phi_1 f_0\,\mathrm{d}\boldsymbol{c} = 0,$$

and

$$\int_{-\infty}^{\infty} c'^2\Phi_1 f_0\,\mathrm{d}\boldsymbol{c} = 0. \tag{3.52}$$

It may further be shown that Φ_1 must have the form

$$\Phi_1 = \left\{-\frac{1}{n}\; A\boldsymbol{c}' \cdot \frac{\partial}{\partial \boldsymbol{r}}(\ln T) + B\boldsymbol{c}'^0\boldsymbol{c}' : \frac{\partial \boldsymbol{c}_0}{\partial \boldsymbol{r}}\right\} \tag{3.53}$$

where A and B are functions of T and c'. The superscript ^{0} above a tensor indicates that the sum of the diagonal components is zero. In the component or subscript notation of Chapter 1,

$$c_i'^0 c_j' = c_i' c_j' - \tfrac{1}{3}c'^2\delta_{ij}.$$

The double product of the two tensors is a scalar quantity which may be written in the subscript notation as

$$c_i'^0 c_j' \frac{\partial c_{0i}}{\partial x_j},$$

or in full Cartesian component form as

$$\left(u'^2-\frac{1}{3}c'^2\right)\frac{\partial u_0}{\partial x}+u'v'\frac{\partial u_0}{\partial y}+u'w'\frac{\partial u_0}{\partial z}+$$

$$+v'u'\frac{\partial v_0}{\partial x}+\left(v'^2-\frac{1}{3}c'^2\right)\frac{\partial v_0}{\partial y}+v'w'\frac{\partial v_0}{\partial z}+$$

$$+w'u'\frac{\partial w_0}{\partial x}+w'v'\frac{\partial w_0}{\partial y}+\left(w'^2-\frac{1}{3}c'^2\right)\frac{\partial w_0}{\partial z},$$

where $c'^2 = u'^2+v'^2+w'^2$.

The substitution of eqn (3.53) into eqns (1.20) and (1.27), through eqns (3.51) and (3.3), provides expressions for the shear stress tensor and heat flux vector as linear functions of the velocity and temperature gradients, respectively. The coefficients of the gradients may be identified with the familiar coefficients of viscosity and heat conduction. The substitution of these relationships for the shear stress tensor and heat flux vector into the conservation equations leads to the *Navier–Stokes equations* of continuum gas dynamics. Since Φ_1 must be small compared with unity, eqn (3.53) shows that there is a limit to the magnitude of the gradients for which the Chapman–Enskog method is valid. We will find that the condition for validity is that the mean free path is small compared with the scale length of the gradients or, less precisely, that the Knudsen number is small with the gas not too far from equilibrium. The evaluation of the next term in the series defined by eqn (3.51) leads to a set of very complicated higher order equations, called the *Burnett equations*. However, the weight of evidence indicates that these equations cannot be relied upon to improve the description that is provided by the Navier–Stokes equations. Therefore, while the Chapman–Enskog method provides analytical expressions for the transport properties and is an essential link between the macroscopic and microscopic approaches, it has not led to the solution of problems that lie outside the range of validity of the Navier–Stokes equations.

The Chapman–Enskog method is characterised by extreme mathematical complexity and the reader is referred to Chapman and Cowling (1952), Hirschfelder, Curtiss, and Bird (1954), Vincenti and Kruger (1965), Harris (1971), or Cercignani (1969). The final results for the transport coefficients for a number of molecular models are presented in Section 4.4. The velocity distribution functions corresponding to these results also constitute important reference quantities and are also discussed in § 4.4. The molecular model enters in the evaluation of the coefficients A and B of eqn (3.53). These are usually obtained as series of Sonine polynomials. Since the word 'order' has already been applied to the specification of the number of terms in eqn (3.51), the solution involving n terms of the Sonine polynomial series is usually referred to as the 'nth approximation'.

Exercises

3.1. Write down the collision term of the Boltzmann equation for the special cases of inverse power law, Maxwell, and hard sphere molecules.

3.2. Show that, if the equation of state is taken into account, there are thirteen dependent variables in the conservation equations.

3.3. A generalized Boltzmann H-function may be based on the distribution function in phase space, rather than velocity space, i.e.

$$H = \int_{-\infty}^{\infty}\int_{-\infty}^{\infty} nf_0 \ln (nf_0)\, \mathrm{d}\boldsymbol{c}\, \mathrm{d}\boldsymbol{r}.$$

Show that, in a equilibrium gas, this function is given by

$$H_0 = -S/k,$$

where S is the absolute entropy.

3.4. Verify that the viscous stress tensor $\boldsymbol{\tau}$ and the heat flux vector $\boldsymbol{q}$ are zero in an equilibrium gas.

3.5. Show that, for Maxwell molecules, the collision integral $\Delta[uv]$ is given by the expression for $\Delta[u^2]$ (eqn 3.38) with τ_{xx} replaced by τ_{xy}.

4

REFERENCE STATES AND BOUNDARY CONDITIONS

4.1. Spatial quantities in an equilibrium gas

THE results that were obtained in § 3.4 for the equilibrium distribution function f_0 may be summarized as follows:

$$f_0 = (\beta^3/\pi^{\frac{3}{2}}) \exp(-\beta^2 c'^2),$$

where (4.1)

$$\beta = (2RT)^{-\frac{1}{2}} = \{m/(2kT)\}^{\frac{1}{2}}.$$

The fraction of the molecules that are located within a velocity space element of volume $\mathrm{d}\boldsymbol{c}$ and located at $\boldsymbol{c}'$ follows from eqn (3.1) as

$$\frac{\mathrm{d}N}{N} = \frac{\mathrm{d}n}{n} = (\beta^3/\pi^{\frac{3}{2}}) \exp(-\beta^2 c'^2)\, \mathrm{d}\boldsymbol{c}. \tag{4.2}$$

The peculiar or thermal velocity $\boldsymbol{c}' = \boldsymbol{c} - \boldsymbol{c}_0$ so that, in Cartesian coordinates (u, v, w), the fraction of molecules with velocity components between u and $u + \mathrm{d}u$, v to $v + \mathrm{d}v$, and w to $w + \mathrm{d}w$ is

$$\frac{\mathrm{d}n}{n} = (\beta^3/\pi^{\frac{3}{2}}) \exp[-\beta^2\{(u-u_0)^2 + (v-v_0)^2 + (w-w_0)^2\}]\, \mathrm{d}u\, \mathrm{d}v\, \mathrm{d}w. \tag{4.3}$$

For polar coordinates (c', θ, ϕ) in frame of reference moving with the stream velocity, the volume of the velocity space element is $c'^2 \sin\theta\, \mathrm{d}\theta\, \mathrm{d}\phi\, \mathrm{d}c'$. The number of molecules with speed between c' and $c' + \mathrm{d}c'$, which make an angle between θ and $\theta + \mathrm{d}\theta$ with the polar direction, and have an azimuth angle between ϕ and $\phi + \mathrm{d}\phi$ is, therefore,

$$\frac{\mathrm{d}n}{n} = (\beta^3/\pi^{\frac{3}{2}}) c'^2 \exp(-\beta^2 c'^2) \sin\theta\, \mathrm{d}\theta\, \mathrm{d}\phi\, \mathrm{d}c'. \tag{4.4}$$

The fraction of molecules with speed between c' and $c' + \mathrm{d}c'$, irrespective of direction, is obtained from eqn (4.4) by integrating ϕ between the limits 0 to 2π and θ from 0 to π, to give

$$(4/\pi^{\frac{1}{2}}) \beta^3 c'^2 \exp(-\beta^2 c'^2)\, \mathrm{d}c'. \tag{4.5}$$

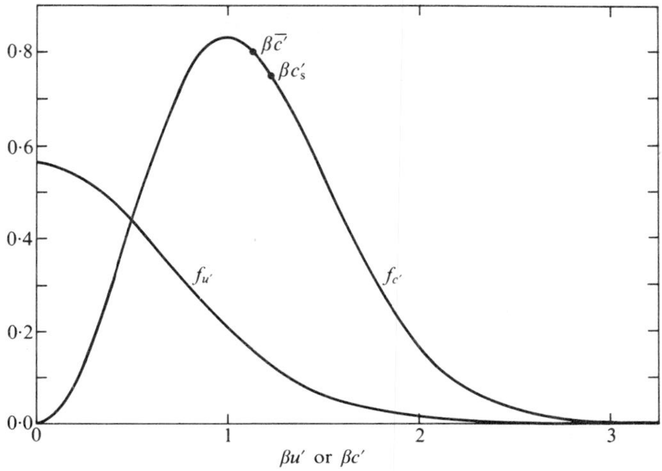

FIG. 4.1. Equilibrium distribution functions for the molecular speed and for a molecular velocity component.

A distribution function $f_{c'}$ may be defined such that the fraction of molecules with speeds between c' and $c'+\mathrm{d}c'$ if $f_{c'}\,\mathrm{d}c'$. Eqn (4.5) shows that

$$f_{c'} = (4/\pi^{\frac{1}{2}})\beta^3 c'^2 \exp(-\beta^2 c'^2). \tag{4.6}$$

The function $f_{c'}$ is plotted in Fig. (4.1). It is zero when c' is zero, increases to a maximum when $\beta c'$ is unity, and then decreases as c' increases. The parameter β is, therefore, the reciprocal of the *most probable molecular thermal speed* c'_m i.e.

$$c'_m = 1/\beta. \tag{4.7}$$

The average of any quantity depending on the molecular speed may be obtained through the application of eqn (3.3) to the thermal speed distribution function $f_{c'}$. The *average thermal speed* $\overline{c'}$ is

$$\overline{c'} = \int_0^\infty c' f_{c'}\,\mathrm{d}c' = (4/\pi^{\frac{1}{2}})\beta^3 \int_0^\infty c'^3 \exp(-\beta^2 c'^2)\,\mathrm{d}c'$$

or, referring to the standard integrals in Appendix B,

$$\overline{c'} = 2/(\pi^{\frac{1}{2}}\beta) = (2/\pi^{\frac{1}{2}})c'_m. \tag{4.8}$$

The mean square thermal velocity has already been determined in eqn (3.47) through the evaluation of β in terms of T. The *root mean square thermal speed* c'_s follows as

$$c'_s = (\tfrac{3}{2})^{\frac{1}{2}}(1/\beta) = (\tfrac{3}{2})^{\frac{1}{2}}c'_m = (\tfrac{3}{8}\pi)^{\frac{1}{2}}\overline{c'}. \tag{4.9}$$

The ordering $c'_s > \overline{c'} > c'_m$ is a consequence of the high speed tail of the distribution function. The fraction of molecules with thermal speed above some value c' is given by the integral of eqn (4.5) from c' to ∞, i.e.

$$(4/\pi^{\frac{1}{2}})\beta^3 \int_{c'}^{\infty} c'^2 \exp(-\beta^2 c'^2)\,\mathrm{d}c'.$$

Again using the standard integrals of Appendix B, this fraction becomes

$$1+(2/\pi^{\frac{1}{2}})\beta c' \exp(-\beta^2 c'^2) - \operatorname{erf}(\beta c'). \tag{4.10}$$

The error function

$$\operatorname{erf}(x) = (2/\pi^{\frac{1}{2}}) \int_0^x \exp(-x^2)\,\mathrm{d}x \tag{4.11}$$

will be encountered frequently. Appendix C presents tabulated values of this function, together with its limiting values, an asymptotic series, and a series that is readily programmed.

The fraction of molecules with a velocity component within a given range, irrespective of the magnitude of the other components, is obtained by integrating eqn (4.2) over these other components. For example, the fraction of molecules with a thermal velocity component in the x direction of between u' and $u'+\mathrm{d}u'$ is

$$(\beta^3/\pi^{\frac{3}{2}}) \int_{-\infty}^{\infty} \int_{-\infty}^{\infty} \exp\{-\beta^2(u'^2+v'^2+w'^2)\}\,\mathrm{d}u'\,\mathrm{d}v'\,\mathrm{d}w',$$

or

$$(\beta/\pi^{\frac{1}{2}}) \exp(-\beta^2 u'^2)\,\mathrm{d}u. \tag{4.12}$$

The distribution function for a thermal velocity component is, therefore,

$$f_{u'} = (\beta/\pi^{\frac{1}{2}}) \exp(-\beta^2 u'^2). \tag{4.13}$$

This function is also plotted in Fig. 4.1. Since the velocity distribution function f is spherically symmetric in velocity space about the point representing the stream velocity, the most probable value of a particular thermal velocity component is zero. The average of the thermal velocity components in the x direction averaged over only those molecules moving in the positive x direction is

$$\begin{aligned} \int_0^{\infty} u' f_{u'}\,\mathrm{d}u' \Big/ \int_0^{\infty} f_{u'}\,\mathrm{d}u' &= 2(\beta/\pi^{\frac{1}{2}}) \int_0^{\infty} u' \exp(-\beta^2 u'^2)\,\mathrm{d}u' \\ &= 1/(\pi^{\frac{1}{2}}\beta) \end{aligned} \tag{4.14}$$

or, comparing this result with eqn (4.8), $\overline{c'}/2$.

For gas mixtures in equilibrium, the equations of this section may be applied separately to each component.

4.2. Fluxal quantities in an equilibrium gas

We will now consider the flux of molecular quantities across a surface element in an equilibrium gas. The stream velocity $\boldsymbol{c}_0$ is inclined at the angle θ to the unit normal vector $\boldsymbol{e}$ to the surface element, as shown in Fig. 4.2.

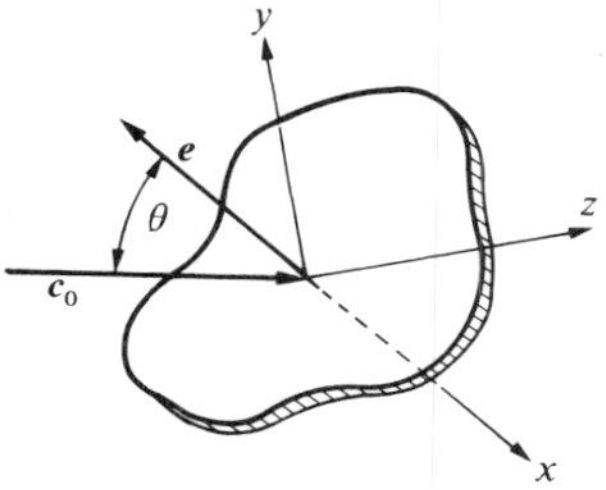

FIG. 4.2. Coordinate system for the analysis of the molecular flux across a surface element.

Without loss of generality, we may choose Cartesian coordinates such that the stream velocity lies in the x, y plane and the surface element lies in the y, z-plane, with the x-axis in the negative $\boldsymbol{e}$ direction. Each molecule has velocity components

$$u = u' + c_0 \cos\theta,$$
$$v = v' + c_0 \sin\theta, \tag{4.15}$$

and

$$w = w'.$$

Therefore, from eqn (1.12), the *inward* (i.e. in the negative $\boldsymbol{e}$ or positive x direction) flux of some quantity Q is

$$n\overline{Qu},$$

or

$$n\int_{-\infty}^{\infty}\int_{-\infty}^{\infty}\int_{0}^{\infty} Quf \,\mathrm{d}u\,\mathrm{d}v\,\mathrm{d}w, \tag{4.16}$$

where consideration has been limited to those molecules moving in the positive x direction. For an equilibrium gas, the function f_0 may be substituted from eqn (4.1) to give the inward flux of the quantity Q across the element as

$$\frac{n\beta^3}{\pi^{\frac{3}{2}}}\int_{-\infty}^{\infty}\int_{-\infty}^{\infty}\int_{0}^{\infty} Qu\exp\{-\beta^2(u'^2+v'^2+w'^2)\}\,du\,dv\,dw$$

per unit area per unit time. Eqn (4.12) enables this result to be written in terms of the stream velocity and the thermal velocity components only i.e.

$$\frac{n\beta^3}{\pi^{\frac{3}{2}}}\int_{-\infty}^{\infty}\int_{-\infty}^{\infty}\int_{-c_0\cos\theta}^{\infty} Q(u'+c_0\cos\theta)\times \times\exp\{-\beta^2(u'^2+v'^2+w'^2)\}\,du'\,dv'\,dw'. \tag{4.17}$$

The *inward number flux* N_i to the element is obtained by setting $Q = 1$ in eqn (4.17). The variables in the multiple integral may be separated to give

$$N_i = \frac{n\beta^3}{\pi^{\frac{3}{2}}}\int_{-\infty}^{\infty}\exp(-\beta^2 w'^2)\,dw'\int_{-\infty}^{\infty}\exp(-\beta^2 v'^2)\,dv'\times \times\int_{-c_0\cos\theta}^{\infty}(u'+c_0\cos\theta)\exp(-\beta^2u'^2)\,du'.$$

The standard integrals of Appendix B enable this to be written

$$\beta N_i/n = [\exp(-s^2\cos^2\theta)+\pi^{\frac{1}{2}}s\cos\theta\{1+\operatorname{erf}(s\cos\theta)\}]/(2\pi^{\frac{1}{2}}), \tag{4.18}$$

where

$$s = c_0\beta = c_0/c'_m = c_0/(2RT)^{\frac{1}{2}} \tag{4.19}$$

is called the *molecular speed ratio.*

For a stationary gas ($s = c_0 = 0$), this reduces to

$$\beta N_i/n = 1/(2\pi^{\frac{1}{2}})$$

or, from eqn (4.8) and noting that $c = c'$,

$$N_i = \tfrac{1}{4}n\bar{c}. \tag{4.20}$$

This result could have been deduced by physical reasoning from the stationary gas version of eqn (4.14). This states that the mean velocity component, taken over the molecules for which it is positive, is equal to half the mean molecular speed $\bar{c}$. Since half the molecules have a positive component in a particular direction, the number flux per unit area per unit volume in a particular direction must be $\frac{1}{4}n\bar{c}$.

The *inward normal momentum flux* p_i to the element is obtained by setting $Q = mu = m(u' + c_0 \cos\theta)$ in eqn (4.17) to give

$$p_i = \frac{nm\beta^3}{\pi^{\frac{3}{2}}} \int_{-\infty}^{\infty} \exp(-\beta^2 w'^2)\,dw' \int_{-\infty}^{\infty} \exp(-\beta^2 v'^2)\,dv' \times$$
$$\times \int_{-c_0\cos\theta}^{\infty} (u' + c_0 \cos\theta)^2 \exp(-\beta^2 u'^2)\,du',$$

or

$$\beta^2 p_i/\rho = [s\cos\theta \exp(-s^2\cos^2\theta) + \pi^{\frac{1}{2}}\{1 + \operatorname{erf}(s\cos\theta)\} \times \times(\tfrac{1}{2} + s^2\cos^2\theta)]/(2\pi^{\frac{1}{2}}). \tag{4.21}$$

For a stationary gas, this gives,

$$p_i = \rho/(4\beta^2) = \rho RT/2 = p/2,$$

as would be expected in an equilibrium gas since the inward moving molecules contribute half the pressure. Similarly, the *inward parallel momentum flux* τ_i is obtained for $Q = mv = m(v' + c_0 \sin\theta)$. The integral is

$$\tau_i = (nm\beta^3/\pi^{\frac{3}{2}}) \int_{-\infty}^{\infty} \exp(-\beta^2 w'^2)\,dw' \int_{-\infty}^{\infty} (v' + c_0\sin\theta) \exp(-\beta^2 v'^2)\,dv'$$
$$\times \int_{-c_0\cos\theta}^{\infty} (u' + c_0\cos\theta) \exp(-\beta^2 u'^2)\,du'$$

and the final result follows as,

$$\beta^2 \tau_i/\rho = s\sin\theta\,[\exp(-s^2\cos^2\theta) + \pi^{\frac{1}{2}} s\cos\theta\{1 + \operatorname{erf}(s\cos\theta)\}]/(2\pi^{\frac{1}{2}})$$
$$= s\sin\theta\,(\beta N_i/n). \tag{4.22}$$

The parallel momentum flux in a stationary equilibrium gas is, of course, zero as a consequence of the symmetry of the distribution function.

Finally the *inward translational energy flux* $q_{i,tr}$ to the element is obtained by setting $Q = \frac{1}{2}mc^2 = \frac{1}{2}m(u^2 + v^2 + w^2)$ in eqn (4.17) to give

$$q_{i'tr} = (nm\beta^3/2\pi^{\frac{3}{2}}) \int_{-\infty}^{\infty} \int_{-\infty}^{\infty} \int_{-c_0\cos\theta}^{\infty} \{(u' + c_0\cos\theta)^2 + (v' + c_0\sin\theta)^2 +$$
$$+ w'^2\}(u' + c_0\cos\theta) \exp\{-\beta^2(u'^2 + v'^2 + w'^2)\}\,du'\,dv'\,dw'$$

or

$$\beta^3 q_{i,tr}/\rho = [(s^2 + 2)\exp(-s^2\cos^2\theta) + \pi^{\frac{1}{2}} s\cos\theta(s^2 + 5/2) \times \times\{1 + \operatorname{erf}(s\cos\theta)\}]/(4\pi^{\frac{1}{2}}). \tag{4.23}$$

For a stationary gas this becomes

$$\beta^3 q_{i,tr}/\rho = 1/(2\pi^{\frac{1}{2}}). \tag{4.24}$$

Unlike the number and momentum flux, the energy flux is modified by the presence of internal energy. For an equilibrium non-reacting gas, the specific internal energy follows from eqn (1.25) as

$$e_{int} = \tfrac{1}{2}\zeta RT.$$

Here, ζ is the number of internal degrees of freedom and may be related to the specific heat ratio of the gas by

$$\gamma = (\zeta+5)/(\zeta+3),$$

so that

$$\zeta = (5-3\gamma)/(\gamma-1). \tag{4.25}$$

The inward *internal energy flux* is equal to the product of the number flux N_i and the average internal energy per molecule. Therefore, from eqns (1.25), (4.18), and (4.25),

$$\beta^3 q_{i,int}/\rho = \{(5-3\gamma)/(\gamma-1)\}[\exp(-s^2\cos^2\theta)+ \\ +\pi^{\frac{1}{2}}s\cos\theta\{1+\operatorname{erf}(s\cos\theta)\}]/(8\pi^{\frac{1}{2}}). \tag{4.26}$$

The *total energy flux* to the element is obtained by summing eqn (4.23) for $q_{i,tr}$ and eqn (4.26) for $q_{i,int}$ i.e.:

$$\beta^3 q_i/\rho = [\{2s^2+(\gamma+1)/(\gamma-1)\}\exp(-s^2\cos^2\theta)+2\pi^{\frac{1}{2}}s\cos\theta\{s^2+\gamma/(\gamma-1)\}\times \\ \times\{1+\operatorname{erf}(s\cos\theta)\}]/(8\pi^{\frac{1}{2}}). \tag{4.27}$$

For a stationary gas, this becomes

$$\beta^3 q_i/\rho = \{(\gamma+1)/(\gamma-1)\}/(8\pi^{\frac{1}{2}}). \tag{4.28}$$

The average energy of the molecules crossing the surface element in a stationary monatomic gas is given by q_i/N_i for $s = 0$, or m/β^2. The average energy of a molecule in a spatial element is given by eqn (1.23) as $\frac{3}{2}mRT$ or $\frac{3}{4}m/\beta^2$. The average energy of the molecules crossing a surface element therefore exceeds that in a spatial element by the factor 4/3. The physical explanation for this is that the probability of a fast molecule crossing a surface element in a given time is greater than the corresponding probability for a slower molecule. This is the reason why the total energy flux of eqn (4.28) is exactly twice the translational energy flux of eqn (4.24) when $\zeta = 4$, rather than when $\zeta = 3$.

The above equation for N_i is usually derived in the context of molecular effusion in a rarefied gas. This problem is dealt with in detail in the following chapter. The other equations for p_i, τ_i, and q_i are generally introduced in

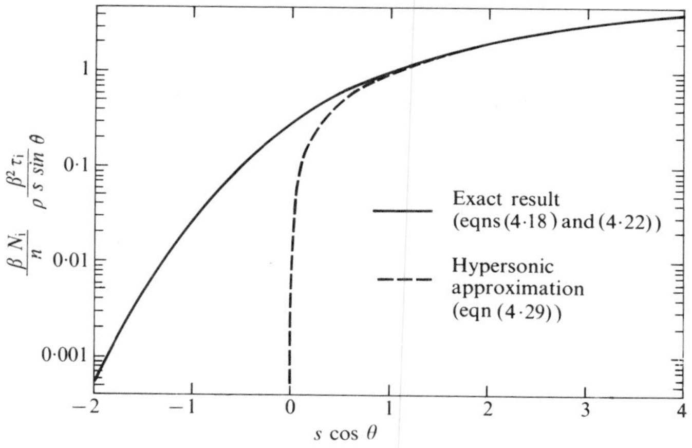

FIG. 4.3. Inward number and parallel momentum flux across a surface element.

the context of free-molecule aerodynamics. They then represent the pressure, shear stress, and heat flux incident on an element of solid surface. However, it must be emphasized that the equations apply to the flux quantities across a surface element in an equilibrium gas at any density and constitute important reference quantities.

The general behaviour of the molecular number flux in an equilibrium gas is illustrated in Fig. 4.3. As $s \cos \theta \to \infty$, $\exp(-s^2 \cos^2 \theta) \to 0$, and $\text{erf}(s \cos \theta) \to 1$ so that the hypersonic form of eqn (4.18) is

$$\beta N_i/n = s \cos \theta,$$

or

$$N_i = nc_0 \cos \theta. \tag{4.29}$$

This is the equation that would be derived from first principles if one neglects the thermal velocity components and considers only the effect due to the stream speed. Eqn (4.22) shows that $\beta^2 \tau_i/(\rho s \sin \theta)$ is equal to $\beta N_i/n$, so that the behaviour of the parallel momentum flux is also covered by Fig. 4.3. The effect of the thermal velocity components is small at $s \cos \theta = 1$ where $\beta N_i/n = 1{\cdot}02513$, and almost negligible at $s \cos \theta = 2$ where $\beta N_i/n = 2{\cdot}00049$.

A similar plot for the non-dimensional normal momentum flux $\beta^2 p_i/\rho$ is presented in Fig. 4.4. Eqn (4.21) shows that the limiting value of this quantity as the exponential and error function terms tend to zero and unity, respectively, is

$$\beta^2 p_i/\rho = \tfrac{1}{2} + s^2 \cos^2 \theta,$$

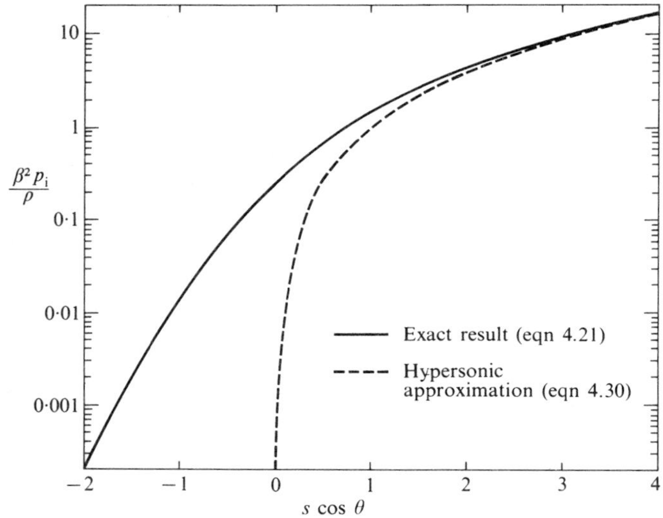

FIG. 4.4. Inward normal momentum flux across a surface element.

or

$$p_i/(\tfrac{1}{2}\rho c_0^2) = 1/s^2 + 2\cos^2\theta. \tag{4.30}$$

The departure of the exact result from this is again very small for $s\cos\theta > 2$. In the context of free molecule aerodynamics, the incident pressure coefficient tends to its limiting hypersonic value of $2\cos^2\theta$ only as s becomes very large in comparison with unity. The additional term $1/s^2$ in eqn (4.30) represents the enhanced normal momentum flux due to the thermal velocity and means that the hypersonic limit is approached much less rapidly than in the case of the number flux.

The energy flux is much more complex since it depends on s and θ independently, rather than on $s\cos\theta$, and also depends on the number of internal degrees of freedom. The limiting value of eqn (4.27) is

$$\beta^3 q_i/\rho = \tfrac{1}{2}s\cos\theta\{s^2 + \gamma/(\gamma-1)\}$$

or

$$q_i/(\tfrac{1}{2}\rho c_0^3) = \cos\theta[1 + \gamma/\{s^2(\gamma-1)\}]. \tag{4.31}$$

The hypersonic limit of $\tfrac{1}{2}\rho c_0^3\cos\theta$ for q_i is again approached as $1/s^2$ becomes small, although the additional factor of $\gamma/(\gamma-1)$ means that the approach is less rapid. The behaviour of the energy flux as functions of s, θ, and $s\cos\theta$ is illustrated in Fig. 4.5. This brings out the very strong dependence of the

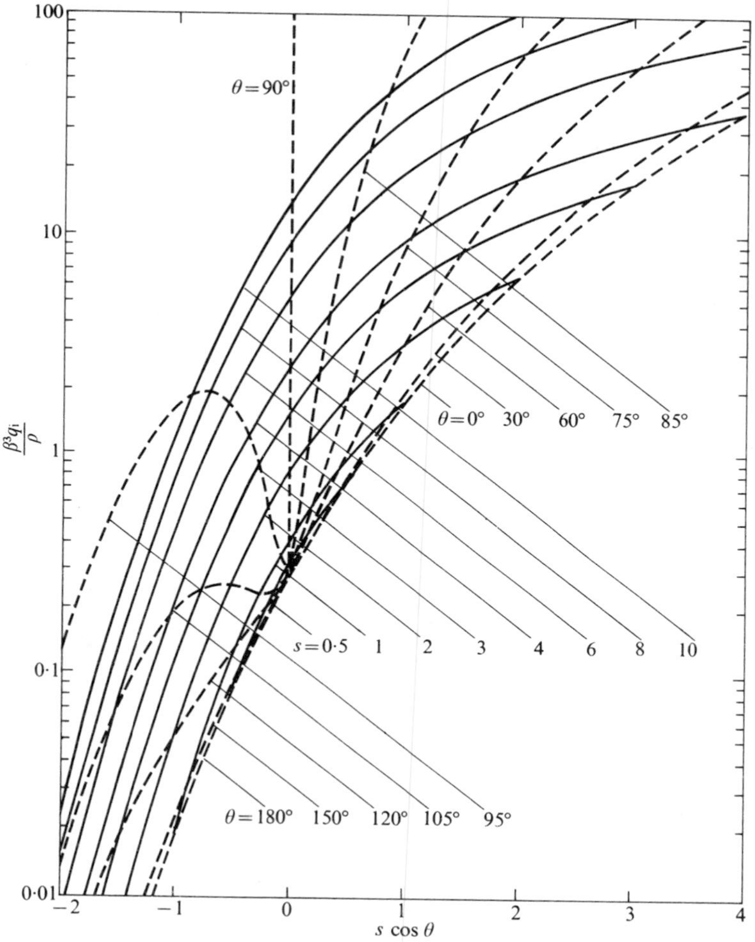

FIG. 4.5. Inward energy flux across a surface element in a monatomic gas.

heat transfer on the inclination of the element when it is almost aligned with the flow direction (i.e. when θ is near 90°).

4.3. Collisional quantities in an equilibrium gas

General expressions were obtained in Chapter 1 for the mean collision rate and mean free path in a dilute gas. These involve the mean value of the product of the collision cross-section and the magnitude of the relative velocity. Eqn (2.27) shows that the total collision cross-section of an inverse power law molecule is proportional to the relative speed raised to the power $-4/(\eta-1)$. Special cases include the hard-sphere molecule ($\eta = \infty$) with a fixed cross-section and Maxwell molecules ($\eta = 5$). The mean collision rate is therefore proportional to the mean value of $c_r^{(\eta-5)/(\eta-1)}$. This will now be evaluated for an equilibrium gas. Because an arbitrary distance or deflection angle cut-off must be applied to the general power law molecule, the final results are generally presented for the hard-sphere case only. However, the functional dependence of the intermediate results, particularly the temperature dependence, is important and the basic analysis will be based on the more general model.

The relative velocity in a binary collision is $\boldsymbol{c}_r = \boldsymbol{c}_1 - \boldsymbol{c}_2$, where the subscripts 1 and 2 denote the two molecules that are involved. Assuming molecular chaos, the two-particle distribution function is equal to the product of the two single-particle distribution functions f_1 and f_2. The required mean value is then

$$\overline{c_r^{(\eta-5)/(\eta-1)}} = \int_{-\infty}^{\infty}\int_{-\infty}^{\infty} c_r^{(\eta-5)/(\eta-1)} f_1 f_2 \, d\boldsymbol{c}_1 \, d\boldsymbol{c}_2.$$

and, from eqn (4.1) for an equilibrium gas,

$$\overline{c_r^{(\eta-5)/(\eta-1)}} = \frac{(m_1 m_2)^{\frac{3}{2}}}{(2\pi k T)^3}\int_{-\infty}^{\infty}\int_{-\infty}^{\infty} c_r^{(\eta-5)/(\eta-1)} \exp\left\{\frac{-(m_1 c_1^2 + m_2 c_2^2)}{2kT}\right\} d\boldsymbol{c}_1 \, d\boldsymbol{c}_2.$$

This integral is most easily evaluated if the variables are changed from $\boldsymbol{c}_1$ and $\boldsymbol{c}_2$ to $\boldsymbol{c}_r$ and $\boldsymbol{c}_m$. The Jacobean of the transformation is

$$\frac{\partial(u_1, v_1, w_1, u_2, v_2, w_2)}{\partial(u_r, v_r, w_r, u_m, v_m, w_m)}$$

but, because of the symmetry of eqns (2.3) and (2.4), it is sufficient to evaluate the reduced one-dimensional Jacobean. This is, in the x direction,

$$\frac{\partial(u_1, u_2)}{\partial(u_r, u_m)} = \begin{vmatrix} \dfrac{\partial u_1}{\partial u_r} & \dfrac{\partial u_1}{\partial u_m} \\[2ex] \dfrac{\partial u_2}{\partial u_r} & \dfrac{\partial u_2}{\partial u_m} \end{vmatrix}$$

or, using eqn (2.4),

$$\begin{vmatrix} \dfrac{m_2}{m_1+m_2} & 1 \\[2ex] -\dfrac{m_1}{m_1+m_2} & 1 \end{vmatrix} = \frac{m_2}{m_1+m_2} + \frac{m_1}{m_1+m_2} = 1.$$

The full Jacobean is therefore unity and $d\boldsymbol{c}_m\, d\boldsymbol{c}_r$ may be substituted for $d\boldsymbol{c}_1\, d\boldsymbol{c}_2$ in the above integral. We also have, from eqn (2.6)

$$m_1 c_1^2 + m_2 c_2^2 = (m_1 + m_2) c_m^2 + m_r c_r^2,$$

where m_r is the reduced mass. Since the integrand is independent of the directions of $\boldsymbol{c}_r$ and $\boldsymbol{c}_m$, the velocity space elements may be written in polar coordinates and an integration made over all polar and azimuth angles to give

$$d\boldsymbol{c}_r = 4\pi c_r^2\, dc_r \quad \text{and} \quad d\boldsymbol{c}_m = 4\pi c_m^2\, dc_m.$$

Therefore,

$$\overline{c_r^{(\eta-5)/(\eta-1)}} = \frac{2(m_1 m_2)^{\frac{3}{2}}}{\pi (kT)^3} \int_0^\infty \int_0^\infty c_r^{(3\eta-7)/(\eta-1)} c_m^2 \times$$
$$\times \exp\left[-\{(m_1+m_2)c_m^2 + m_r c_r^2\}/(2kT)\right] dc_m\, dc_r,$$

or

$$\overline{c_r^{(\eta-5)/(\eta-1)}} = \frac{2(m_1 m_2)^{\frac{3}{2}}}{\pi (kT)^3} \int_0^\infty c_m^2 \exp\{-(m_1+m_2)c_m^2/(2kT)\}\, dc_m \times$$
$$\times \int_0^\infty c_r^{(3\eta-7)/(\eta-1)} \exp\{-m_r c_r^2/(2kT)\}\, dc_r. \tag{4.32}$$

Note that the distribution functions for c_m and c_r are

$$f_{c_m} = \frac{4(m_1+m_2)^{\frac{3}{2}}}{\pi^{\frac{1}{2}}(2kT)^{\frac{3}{2}}} c_m^2 \exp\{-(m_1+m_2)c_m^2/(2kT)\}$$

and (4.33)

$$f_{c_r} = \frac{4 m_r^{\frac{3}{2}}}{\pi^{\frac{1}{2}}(2kT)^{\frac{3}{2}}} c_r^2 \exp\{-m_r c_r^2/(2kT)\}.$$

The application of the standard integrals of Appendix B to eqn (4.32) gives

$$\overline{c_r^{(\eta-5)/(\eta-1)}} = (2/\pi^{\frac{1}{2}})\Gamma\{2(\eta-2)/(\eta-1)\}(2kT/m_r)^{(\eta-5)/\{2(\eta-1)\}}. \tag{4.34}$$

For the special case of a hard-sphere gas, the result reduces to

$$\overline{c_r} = (2/\pi^{\frac{1}{2}})(2kT/m_r)^{\frac{1}{2}}.$$

For a simple gas, $m_r = m/2$ and the result is

$$\overline{c_r} = 2^{\frac{3}{2}}/(\pi^{\frac{1}{2}}\beta) = 2^{\frac{1}{2}}\overline{c'}. \tag{4.35}$$

The mean collision rate per molecule in an equilibrium simple gas of hard-sphere molecules is given by the substitution of eqn (4.35) into eqn (1.6*a*), i.e.

$$\nu_0 = 2^{\frac{1}{2}}\pi d^2 n\overline{c'}. \tag{4.36}$$

The number of collisions per unit time per unit volume follows from eqn (1.7) as

$$N_{c_0} = 2^{-\frac{1}{2}}\pi d^2 n^2 \overline{c'}. \tag{4.37}$$

The mean free path in an equilibrium simple gas of hard sphere molecules is then given by eqns (4.36) and (1.8) as

$$\lambda_0 = (2^{\frac{1}{2}}\pi d^2 n)^{-1}. \tag{4.38}$$

Note that the mean free path has been defined as the average distance travelled by a molecule between collisions, this leading to the ratio $\overline{c_r}/\overline{c'}$ in eqn (1.8). A different result is obtained through a similar calculation involving the ratio $\overline{c_r/c'}$. This calculates the mean distance travelled by the molecules to their next collision starting from a given instant (rather than from the non-simultaneous previous collisions). This is called the Tait mean free path and leads to the replacement of the $2^{\frac{1}{2}}$ in eqn (4.38) by 1·477.

Now consider a gas mixture consisting of s separate hard-sphere molecular species. Eqns (1.30*a*) and (4.34), with $m_r = m_p m_q/(m_p+m_q)$, give the mean collision rate for a species p molecule with species q molecule as

$$\nu_{pq} = 2\pi^{\frac{1}{2}} d_{pq}^2 n_q \{2kT(m_p+m_q)/m_p m_q\}^{\frac{1}{2}}, \tag{4.39}$$

where $d_{pq} = \frac{1}{2}(d_p+d_q)$. As before, the mean collision rate for species p molecules is

$$\nu_p = \sum_{q=1}^{s} \nu_{pq}, \tag{4.40}$$

and the mean collision rate per molecule for the mixture is

$$\nu = \sum_{p=1}^{s} \{(n_p/n)\nu_p\}. \tag{4.41}$$

The equations for the number of collisions per unit time also remain unaltered from those given in § 1.3. The equilibrium mean free path for species p molecules is, from eqns (1.34), (4.1), and (4.40),

$$\lambda_{p_0} = \frac{\overline{c'_p}}{\nu_p} = \left(\sum_{q=1}^{s}\left[2\pi^{\frac{1}{2}} d_{pq}^2 n_q \left\{\frac{2kT(m_p+m_q)}{m_p m_q}\right\}^{\frac{1}{2}} \Big/ \left\{\frac{2}{\pi^{\frac{1}{2}}}\left(\frac{2kT}{m_p}\right)^{\frac{1}{2}}\right\}\right]\right)^{-1},$$

or

$$\lambda_{po} = \left[\sum_{q=1}^{s}\left\{\pi d_{pq}^2 n_q\left(1+\frac{m_p}{m_q}\right)^{\frac{1}{2}}\right\}\right]^{-1}. \tag{4.42}$$

Finally, the mean free path for the mixture is

$$\lambda_0 = \sum_{p=1}^{s}\left(\frac{n_p}{n}\left[\sum_{q=1}^{s}\left\{\pi d_{pq}^2 n_q\left(1+\frac{m_p}{m_q}\right)^{\frac{1}{2}}\right\}\right]^{-1}\right). \tag{4.43}$$

The mean value, over all collisions, of any quantity Q that is a function of c_r alone may be obtained by including Q within the integral over c_r in eqn (4.32), and then dividing the resulting expression by eqn (4.34), i.e.

$$\bar{Q} = \left[2\left(\frac{m_r}{2kT}\right)^{2(\eta-2)/(\eta-1)}\bigg/\Gamma\left\{\frac{2(\eta-2)}{\eta-1}\right\}\right]\int_0^{\infty} Q c_r^{(3\eta-7)/(\eta-1)}\exp\left(\frac{-m_r c_r^2}{2kT}\right)dc_r. \tag{4.44}$$

The translational energy in the centre of mass frame of reference is $E_t = \frac{1}{2}m_r c_r^2$ and the substitution of $Q = \frac{1}{2}m_r c_r^2$ in eqn (4.44) gives

$$\bar{E}_t = \{2(\eta-2)/(\eta-1)\}kT. \tag{4.45}$$

For the special case of hard-sphere molecules, this reduces to

$$\bar{E}_t = 2kT. \tag{4.46}$$

The fraction of collisions in which $E_t = \frac{1}{2}m_r c_r^2$ exceeds some reference value E_c will also be required for future reference. The total number of collisions is proportional to the integral in eqn (4.32), and the number of these with $E_t > E_c$ is obtained by setting the lower limit for the integration over c_r equal to $(2E_c/m_r)^{\frac{1}{2}}$. The fraction then becomes

$$\frac{dN}{N} = \left[2\left(\frac{m_r}{2kT}\right)^{2(\eta-2)/(\eta-1)}\bigg/\Gamma\left\{\frac{2(\eta-2)}{\eta-1}\right\}\right]\times \\ \times\int_{(2E_c/m_r)^{\frac{1}{2}}}^{\infty} c_r^{(3\eta-7)/(\eta-1)}\exp\left(\frac{-m_r c_r^2}{2kT}\right)dc_r \tag{4.47}$$

or, if the incomplete gamma function is introduced,

$$\frac{dN}{N} = \Gamma\left(\frac{2(\eta-2)}{\eta-1},\frac{E_c}{kT}\right)\bigg/\Gamma\left(\frac{2(\eta-2)}{\eta-1}\right). \tag{4.47a}$$

For hard sphere molecules, eqn (4.47) reduces to

$$\frac{dN}{N} = \frac{2m_r^2}{(2kT)^2}\int_{(2E_c/m_r)^{\frac{1}{2}}}^{\infty} c_r^3\exp\{-m_r c_r^2/(2kT)\}\,dc_r \tag{4.48}$$

or, evaluating the integral,

$$\frac{dN}{N} = \exp(-E_c/kT)\{(E_c/kT)+1\}. \tag{4.49}$$

An important variation of this result occurs when one is interested only in the energy based on the component of the relative velocity along the apse line of the collision or, for hard-sphere molecules, along the line of centres. This means that, for a given c_r, the angle θ_A must be between 0 and $\cos^{-1}\{(2E_c/m_r)^{\frac{1}{2}}/c_r\}$. An analytical result is possible for the hard-sphere model. For this model, the differential cross-section is

$$\sigma\, d\Omega = (d_{12}^2/4)\sin\chi\, d\chi\, d\varepsilon,$$

but, from eqn (2.21),

$$\chi = \pi - 2\theta_A$$

and

$$\sin\chi\, d\chi = -4\sin\theta_A\cos\theta_A\, d\theta_A.$$

The effective total collision cross-section σ_E of the collisions with the required relative translational energy is, therefore,

$$\sigma_E = -2\pi d_{12}^2 \int_0^{\cos^{-1}\{(2E_c/m_r)^{\frac{1}{2}}/c_r\}} \sin\theta_A\cos\theta_A\, d\theta_A,$$

or, since $\sin\theta_A\, d\theta_A = -d(\cos\theta_A)$,

$$\sigma_E = \pi d_{12}^2\left(1-\frac{2E_c}{m_r c_r^2}\right) \tag{4.50}$$

This compares with $\sigma_T = \pi d_{12}^2$ so that, to obtain the required fraction, the factor in parentheses in eqn (4.50) must be incorporated into the integrand of eqn (4.48), i.e.

$$\frac{dN}{N} = \frac{2m_r^2}{(2kT)^2}\int_{(2E_c/m_r)^{\frac{1}{2}}}^{\infty} c_r^3\left(1-\frac{2E_c}{m_r c_r^2}\right)\exp\{-m_r c_r^2/(2kT)\}\, dc_r.$$

This is readily evaluated to give the strikingly simple result

$$\frac{dN}{N} = \exp(-E_c/kT) \tag{4.51}$$

for the fraction of collisions in a hard-sphere gas in which the component of the relative translational energy along the line of centres exceeds the value E_c.

4.4. The Chapman–Enskog gas

The following results from the Chapman–Enskog theory for transport properties in a simple gas provide a means of setting effective molecular diameters and mean free paths in real gases. Using only the first term in the Sonine polynomial expansions for the parameters A and B in eqn (3.53), the 'first approximation' to the coefficient of viscosity in a hard-sphere gas is

$$\mu = \frac{5m}{16d^2}\left(\frac{RT}{\pi}\right)^{\frac{1}{2}}. \tag{4.52}$$

This may be combined with eqn (4.38) for the mean free path in an equilibrium gas to give

$$\mu = (5/16)\rho\lambda_0(2\pi RT)^{\frac{1}{2}} = (5/32)\pi\rho\overline{c'}\lambda_0. \tag{4.53}$$

Earlier approximate theories had given similar results with $\frac{1}{2}$ as the numerical factor in place of $5\pi/32$. The simpler result is still sometimes used in place of eqns (4.52) and (4.53) for the setting of nominal molecular diameters and free paths. The generalization of eqn (4.52) to a gas of inverse power law molecules is

$$\mu = \frac{5m(RT/\pi)^{\frac{1}{2}}(2mRT/\kappa)^{2/(\eta-1)}}{8A_2(\eta)\Gamma\{4-2/(\eta-1)\}}. \tag{4.54}$$

Here, Γ is the gamma function and $A_2(\eta)$ is a numerical factor defined by

$$A_2(\eta) \equiv \int_0^\infty \sin^2\chi W_0\, dW_0$$

and for which Chapman and Cowling (1952) give the following values

η	5	7	9	11	15	21	∞
$A_2(\eta)$	0·436	0·357	0·332	0·319	0·309	0·307	0·333.

The combination of the hard-sphere relationship between μ and λ_0, given by eqn (4.53), with eqn (4.54) provides the most convenient means of relating the constant κ in the inverse power law to a nominal mean free path, i.e.

$$\lambda_0 = \frac{\sqrt{2}(2mRT/\kappa)^{2/(\eta-1)}}{\pi n A_2(\eta)\Gamma\{4-2/(\eta-1)\}}. \tag{4.55}$$

The coefficient of heat conduction K is related to the coefficient of viscosity, to the first approximation, by

$$K = (15/4)R\mu. \tag{4.56}$$

Since we are dealing with a monatomic gas, the specific heat ratio at constant pressure c_p is equal to $\frac{5}{2}R$, and the Prandtl number

$$(Pr) = \mu c_p/K = \tfrac{2}{3}. \tag{4.57}$$

The above results have required the evaluation of the constants A and B in eqn (3.53) for the Chapman–Enskog distribution function. This may now be written

$$f = f_0\left\{1 - \frac{K}{\rho(RT)^2}\left(\frac{c'^2}{5RT} - 1\right)\mathbf{c}' \cdot \frac{\partial T}{\partial \mathbf{r}} - \frac{\mu}{\rho(RT)^2}\mathbf{c}'^0\mathbf{c}' : \frac{\partial \mathbf{c}_0}{\partial \mathbf{r}}\right\}. \tag{4.58}$$

The physical implications of this equation are more readily appreciated if we consider the special case with the stream velocity in the x direction and in which gradients exist only in the y direction. The hard-sphere result of eqn (4.53) may be used to write the coefficient of viscosity in terms of the mean free path. Also using eqns (3.47), (4.19), and (4.56), eqn (4.58) becomes

$$f = f_0\left\{1 - \frac{15}{16}\pi^{\frac{1}{2}}\beta v'\left(\beta^2 c'^2 - \frac{5}{2}\right)\frac{\lambda}{T}\frac{\partial T}{\partial y} - \frac{5}{4}\pi^{\frac{1}{2}}\beta^2 u'v'\frac{s\lambda}{u_0}\frac{\partial u_0}{\partial y}\right\}. \tag{4.59}$$

The Chapman–Enskog theory requires the two perturbation terms to be small compared with unity. This condition will be violated for sufficiently large thermal velocities, but, since the fraction of these falls as $\exp(-\beta^2 c'^2)$, the overall validity of the distribution may be assessed on the assumption that the thermal speeds are of order $1/\beta$. Eqn (4.59) then shows that the mean free path must be small in comparison with the scale length of the temperature gradient, and that the product of the speed ratio and mean free path must be small in comparison with the scale length of the stream velocity gradient.

Additional terms in the Sonine polynomial expansions may be included to determine higher order approximations. Chapman and Cowling (1958) show that, in the second approximation, the coefficient of viscosity is increased by the factor $3(\eta-5)^2/\{2(\eta-1)(101\eta-113)\}$ and the coefficient of heat conduction by the factor $(\eta-5)^2/\{4(\eta-1)(11\eta-13)\}$. These are simple numerical constant ranging from 0 for Maxwell molecules to 0·0149 and 0·0227 for hard-sphere molecules. However, additional terms are added by the second approximation to the distribution function and the numerical changes to this quantity are much greater. For example, consider a stationary gas with a temperature gradient. The first approximation to the distribution function is given by eqns (3.47) and (4.58) as

$$f = f_0\left\{1 - \frac{4}{5}\left(\beta^2 c'^2 - \frac{5}{2}\right)\frac{K\beta^2}{\rho RT}\mathbf{c}' \cdot \frac{dT}{d\mathbf{r}}\right\}. \tag{4.60}$$

It may be shown that the second approximation is

$$f = f_0\left[1 - \frac{4}{5}\frac{1}{45\eta^2 - 106\eta + 77}\left\{-2(\eta-5)(\eta-1)\beta^4 c'^4 + \right.\right.$$
$$\left.\left. + (59\eta^2 - 190\eta + 147)\beta^2 c'^2 - 10(13\eta^2 - 37\eta + 28)\right\}\frac{K\beta^2}{\rho RT}c' \cdot \frac{dT}{dr}\right], \tag{4.61}$$

where K now represents the second approximation including the above-mentioned numerical factor.

4.5. Gas–surface interactions

A gas in contact with a solid surface is one of the most frequently occurring boundary conditions. Unfortunately, it is also one of the least well understood. Despite a great deal of theoretical and experimental study, the simple models of specular and diffuse reflection that were originally put forward by Maxwell in 1879 remain the most generally useful models for practical applications.

Specular reflection is perfectly elastic with the molecular velocity component normal to the surface being reversed, while that parallel to the surface remains unchanged. This is a useful reference state for analytical studies, but does not occur for real gas–solid combinations. With this model, there is no thermal or viscous boundary layer adjacent to a smooth solid surface.

In *diffuse reflection*, the velocity of each molecule after reflection is independent of its incident velocity. However, the velocities of the reflected molecules as a whole are distributed according to the half-range equilibrium Maxwellian distribution for the molecules that are directed away from the surface. This distribution is for the temperature T_r which may differ from the temperature T_w of the surface. The extent to which the reflected molecules have their temperature adjusted toward that of the surface is indicated by the *accommodation coefficient* a_c, defined by

$$a_c = \frac{q_i - q_r}{q_i - q_w}. \tag{4.62}$$

Here, q_i and q_r are respectively the incident and reflected energy fluxes, while q_w is the energy flux that would be carried away in diffuse reflection with $T_r = T_w$. The range of a_c is from zero for no accommodation to unity for complete thermal accommodation. Accommodation coefficients may also be defined for the normal and tangential components of momentum. However, the accommodation coefficients may be written as functions of the macroscopic pressure, shear stress, and heat flux and, in general, it is preferable to describe the interaction directly in terms of these quantities.

Experiments with engineering surfaces in contact with gases at temperatures of the order of the standard temperature indicate that the reflection process approximates diffuse reflection with complete thermal accommodation. This may be a consequence of such surfaces being microscopically rough with the incident molecules suffering multiple scattering, or of the molecules being momentarily trapped or adsorbed on the surface. On the other hand, experiments with carefully prepared and cleaned surfaces indicate that the accommodation coefficient can be significantly less than one. In fact, for light gases on carefully prepared surfaces, the measured values are small in comparison with unity.

As the translational energy of the incident molecules relative to the surface becomes large in comparison with the value corresponding to the surface temperature, complete thermal accommodation is less readily achieved. Moreover, experiments using molecular beams directed at carefully prepared and cleaned surfaces have shown that, on the average, the reflected molecules retain some momentum parallel to the surface. The greatest practical challenge has been posed by the problem of calculating the drag coefficient of satellites moving through the outer atmosphere. The translational energy of the incident molecules relative to the satellite is approximately 5 eV, compared with the typical 0·025 eV for a molecule of a stationary gas at standard temperature. In addition, the satellite surface is cleaned by outgassing during its long exposure to a very high vacuum. Since the satellite drag coefficient is required for the estimation of atmospheric densities from the observation of satellite orbits, a considerable number of theoretical and experimental studies of gas–surface interactions were initiated in the 1960s.

The theoretical work has been primarily concerned with the interaction of individual incident molecules with simplified models of the surface structure. The theory for the thermal accommodation coefficient has been reviewed by Goodman (1968) and that for the angular scattering by Goodman (1971). The results from the various theories predict very significant departures from the perfectly accommodated diffuse reflection model that has been recommended for use with engineering surfaces. Experimental results for the thermal accommodation coefficient have generally been based on macroscopic heat transfer from heated wires, and those for the scattering angle have been based on molecular beam experiments. In both cases, there is evidence for agreement between theory and experiment. However, in order to obtain repeatable results from the experiments, meticulous attention must be paid to the preparation and cleaning of the surface and to the purity of the gas. This work has not yet led to results that can be used in the context of engineering calculations. This is partly due to the unpredictable effects produced by surface contaminants and partly due to the fact that, even for ideal surfaces and pure monatomic gases, the problem contains eight dimensionless parameters (Barantsev 1972). While a number of empirical reflection models have been proposed (e.g. Nocilla 1963 and Epstein 1967), the inclusion of some arbitrary fraction of specular reflection appears to remain the most practical way of allowing for departures from complete diffuse reflection. However, the choice of this model represents an admission of ignorance and it should not be assumed that the real interaction process will produce a result that lies between the limits set by complete diffuse and complete specular reflection. This point will be brought out in § 5.4 in the analysis of the drag of a sphere in collisionless flow. The *diffuse elastic* model has sometimes been introduced to emphasize this point. This assumes that the molecules are reflected elastically, but in a random direction. It should also be noted that a specular component for a flat surface may become a backscattered component for a rough surface.

A difficulty that is common to most of the theoretical results based on single-particle interaction trajectories and to some of the empirical reflection models is that they do not satisfy the *reciprocity* condition. As applied to the gas–surface interaction, this may be written (Cercignani 1969, Wenaas 1971; or Kuscer 1971).

$$\boldsymbol{c}_{\mathrm{r}} \cdot \boldsymbol{e} P(-\boldsymbol{c}_{\mathrm{r}}, -\boldsymbol{c}_{\mathrm{i}}) \exp(-E_{c_{\mathrm{r}}}/kT) = -\boldsymbol{c}_{\mathrm{i}} \cdot \boldsymbol{e} P(\boldsymbol{c}_{\mathrm{i}}, \boldsymbol{c}_{\mathrm{r}}) \exp(-E_{c_{\mathrm{i}}}/kT). \qquad (4.63)$$

The unit vector $\boldsymbol{e}$ has been taken normal to the surface, which is at the temperature T. $P(\boldsymbol{c}_1, \boldsymbol{c}_2)$ is the probability that a molecule incident on the surface with velocity $\boldsymbol{c}_1$ leaves it with velocity $\boldsymbol{c}_2$, and E_c is the energy of the molecule. This condition is satisfied by both the diffuse and specular models for a gas in equilibrium with a surface. Its proof (Wenaas 1971) for the nonequilibrium situation is based on a model of the surface as a phase space array of a large number of identical components. It is assumed that each of these components interacts only once with a gas molecule; the reciprocity condition then follows as a consequence of the time reversal invariance of the laws of quantum mechanics. There may be some reservation about the validity of the reciprocity principle when the impact energy of the incident molecule is sufficient to produce a collective reaction among the atoms of the surface lattice. Miller and Subbarao (1971) have experimentally verified the reciprocity principle, but their molecular beam experiments were restricted to a narrow energy range and to cases in which the reflected energy was not very different from the incident energy. Cercignani and Lampis (1974) have presented a range of gas–surface interaction models that conform to the reciprocity principle and Kuscer (1974) has developed a consistent set of accommodation coefficients.

Exercises

4.1. Write the equilibrium distribution function in terms of the axial thermal velocity component c'_x and the radial thermal velocity component c'_n. Show that the mean value of c'_n is $\pi^{\frac{1}{2}}/(2\beta)$, and its most probable value is $1/(2^{\frac{1}{2}}\beta)$.

4.2. Prove the following recurrence formula for the averages of powers of the thermal speed in an equilibrium gas;

$$\overline{c'^m} = (\pi/8)(m+1)\overline{c'}^2\ \overline{c'^{(m-2)}},$$

where $m > 2$.

4.3. Find the fraction of molecules in an equilibrium sample of nitrogen at 20 °C that have;

(*i*) thermal speeds in excess of 1000 m s^{-1}, and

(*ii*) thermal velocity components in the x direction in excess of 1000 m s^{-1}. (Ans: (*i*) 0·00935, (*ii*) 0·00035).

4.4. Show that the fraction of molecules in an equilibrium gas that have a velocity component in the x direction between 0 and the most probable molecular speed, but which have a speed greater than the most probable speed is $1/(\pi^{\frac{1}{2}}e)$.

4.5. Consider the flux of molecules across a surface element in a macroscopically stationary equilibrium gas. Show that the number flux of molecules with speeds in the range c to $c+\mathrm{d}c$ and which make an angle between θ to $\theta+\mathrm{d}\theta$ with the normal to the surface is

$$(2/\pi^{\frac{1}{2}})n\beta^3c^3 \exp(-\beta^2c^2)\sin\theta\cos\theta\,\mathrm{d}\theta\,\mathrm{d}c$$

per unit area per unit time.

4.6. Show that the mean speed of the molecules crossing an element of area in an equilibrium gas exceeds the mean speed of the molecules within a volume element by the factor $3\pi/8$.

4.7. Show that the net (i.e. from both sides) number, normal momentum, parallel momentum, and energy fluxes across the surface element of Fig. 4.2 are

$$nc_0\cos\theta,$$

$$\rho RT+\rho c_0^2\cos^2\theta,$$

$$\rho c_0^2\sin\theta\cos\theta,$$

and

$$\rho c_0\cos\theta\{\tfrac{1}{2}c_0^2+a^2/(\gamma-1)\},\ \text{respectively.}$$

4.8. Show that the mean centre of mass speed of collision pairs in a simple gas of hard-sphere molecules is

$$\overline{c_m} = \overline{c'}/2^{\frac{1}{2}} = \overline{c_r}/2.$$

4.9. Show that the collision probability of a molecule moving with thermal speed c' in a simple gas of hard-sphere molecules is proportional to

$$\exp(-\beta^2c'^2)+(\pi^{\frac{1}{2}}/\beta c')(\beta^2c'^2+\tfrac{1}{2})\,\mathrm{erf}(\beta c').$$

4.10. Consider the Chapman–Enskog result for the one-dimensional heat transfer in the x direction in a monatomic gas. The degree of non-equilibrium in this gas may be illustrated through the separate temperatures T^+ and T^- based on the molecules moving in the positive and negative x directions, respectively. Show that, if the mean free path λ is defined by eqn (4.53),

$$\frac{T^- - T^+}{T} = \frac{5}{4}\frac{\lambda}{T}\frac{\mathrm{d}T}{\mathrm{d}x}.$$

4.11. Argon is in equilibrium with a flat solid surface at a temperature of 20 °C. The surface lies in the $y-z$ plane and the gas is on the positive x side of this plane. A molecule with velocity components $u_i = -257\ \mathrm{m\,s^{-1}}$, $v_i = 450\ \mathrm{m\,s^{-1}}$, and $w_i = -284\ \mathrm{m\,s^{-1}}$ is incident on the surface and is reflected with velocity components $u_r = 120\ \mathrm{m\,s^{-1}}$, $v_r = 92\ \mathrm{m\,s^{-1}}$, and $w_r = 138\ \mathrm{m\,s^{-1}}$. According to the reciprocity principle, what is the relative probability of an interaction with incident velocity $-\boldsymbol{c}_r$ and reflected velocity $-\boldsymbol{c}_i$? (Ans: 0·173).

5

COLLISIONLESS FLOWS

5.1. Introduction

IN this chapter we are concerned with flows in the limiting case of very high Knudsen number. This limit may be expressed as

$$(Kn) = \lambda/L \to \infty, \tag{5.1}$$

and the flows must therefore be characterized by very high mean free paths or by very small typical dimensions. The former is normally the case and is a consequence of very low densities. In this limit, collisions between molecules are sufficiently rare in the region of interest that they may be disregarded. The analysis of the flow requires only the specification of the molecular distribution functions at the boundaries, together with the nature of the molecular interactions with any solid surfaces. It is called the *collisionless* or *free-molecule limit*.

In most cases, the specified distribution functions are for an equilibrium gas, although the Chapman–Enskog distribution is occasionally required. The diffuse reflection model is generally used for the representation of solid surfaces, while the specular model is used for idealized solid boundaries. As discussed in § 4.5, the distribution function of the reflected molecules is of the equilibrium form for both these models. When the objective is to determine the forces on solid surfaces of comparatively simple geometry, analytical solutions generally follow directly from the flux equations of § 4.2. These surfaces forces will generally be time independent, even though the flow as a whole may be unsteady. The situation becomes very much more difficult when the molecules that are reflected from one section of the surface may impinge on another part of the surface. This situation occurs primarily in internal flows, but is also encountered in external flows over bodies with concave surfaces. This class of flow is dealt with as a special case in § 5.7.

The determination of the properties of unsteady flow fields is best done through the *collisionless Boltzmann equation* using a transformation introduced by Narasimha (1962). Since intermolecular collisions may be neglected, the collision terms on the right hand side of the Boltzmann equation (3.20) may be set to zero. The resulting equation is

$$\frac{\partial}{\partial t}(nf) + \boldsymbol{c} \cdot \frac{\partial}{\partial \boldsymbol{r}}(nf) = 0, \tag{5.2}$$

and is called the collisionless Boltzmann equation. We have seen from eqn (3.6) that nf is identical with the distribution function in phase space $\mathscr{F}$, so that the functional relationship is $nf(\boldsymbol{c}, \boldsymbol{r}, t)$. We will be concerned with the application of eqn (5.2) to initial value problems for which

$$nf(\boldsymbol{c}, \boldsymbol{r}, t = 0) = n_{\mathrm{i}} f_{\mathrm{i}}(\boldsymbol{c}, \boldsymbol{r}). \tag{5.3}$$

Eqn (5.2) has the form of the Liouville equation and the solution is that nf remains constant along the molecular paths which are the characteristics of the equation, i.e.

$$nf(\boldsymbol{c}, \boldsymbol{r}, t) = n_{\mathrm{i}} f_{\mathrm{i}}(\boldsymbol{c}, \boldsymbol{r} - \boldsymbol{c}t). \tag{5.4}$$

Eqn (5.4) may be multiplied by the molecular quantity Q and integrated over velocity space, to give,

$$n\overline{Q}(\boldsymbol{r}, t) = \int_{-\infty}^{\infty} Q n_{\mathrm{i}} f_{\mathrm{i}}(\boldsymbol{c}, \boldsymbol{r} - \boldsymbol{c}t)\, d\boldsymbol{c}.$$

Now make the transformation from $\boldsymbol{c}$ to

$$\boldsymbol{r}' = \boldsymbol{r} - \boldsymbol{c}t, \tag{5.5}$$

for which the Jacobean is

$$\frac{\partial(x', y', z')}{\partial(u, v, w)} = -t^3.$$

Therefore,

$$n\overline{Q}(\boldsymbol{r}, t) = \frac{1}{t^3} \int Q n_{\mathrm{i}} f_{\mathrm{i}}\left(\frac{\boldsymbol{r} - \boldsymbol{r}'}{t}, \boldsymbol{r}'\right) d\boldsymbol{r}', \tag{5.6}$$

with the limits of the integration being the extent of the gas at time $t = 0$. Typical applications of this equation are described in § 5.3.

The results from collisionless or free molecule theory are important in their own right to the extent that flows at very high Knudsen number occur in practice. They also serve as important reference quantities since they provide the opposite limit to continuum theory which is valid when $(Kn) \ll 1$. A knowledge of the two limits is extremely useful when assessing the results from analysis, numerical study, or experiment in the *transition regime* which lies between the continuum and free-molecule regimes.

The most important general features of free-molecule flows may be illustrated through the very simple example of *molecular effusion*. This is the flow that occurs when an equilibrium gas is separated from a vacuum by a thin wall in which there is a small hole. If the dimensions of the hole are sufficiently small in comparison with the mean free path, each molecule that passes through the hole is subject to a negligibly small probability of having been affected by the flow of other molecules through the hole. The flux through the hole is then the same as that across any small surface element in an equilibrium gas, as analysed in § 4.2. The number flux is given directly by eqn (4.20) and, multiplying this by m, the free molecule mass flux per unit area is

$$\Gamma_f = \frac{nm}{2\pi^{\frac{1}{2}}\beta} = \rho\left(\frac{RT}{2\pi}\right)^{\frac{1}{2}}. \tag{5.7}$$

This may be compared with the result from continuum theory which is valid when the mean free path is very small in comparison with the dimensions of the hole. The continuum result is a choked flow with sonic velocity in the plane of the hole. The continuum mass flux is, therefore,

$$\Gamma_c = \rho^* a^*$$

with the asterisk representing sonic conditions. Using the standard one-dimensional steady continuum flow equations to refer these to the stagnation quantities ρ and T, we have

$$\Gamma_c = \left(\frac{2}{\gamma+1}\right)^{(\gamma+1)/[2(\gamma-1)]} \rho(\gamma RT)^{\frac{1}{2}}. \tag{5.8}$$

Here, γ is the specific heat ratio of the gas so that the continuum and free molecule results differ only by a numerical factor.

A comparison of eqns (5.7) and (5.8) shows that the ratio of the free molecule to the continuum mass flux is

$$\frac{\Gamma_f}{\Gamma_c} = \frac{1}{(2\pi\gamma)^{\frac{1}{2}}}\left(\frac{\gamma+1}{2}\right)^{(\gamma+1)/[2(\gamma-1)]} \tag{5.9}$$
$$= 0{\cdot}5494 \text{ for } \gamma = \tfrac{5}{3}.$$

Assuming a circular hole, the obvious choice of Knudsen number is the ratio of the mean free path in the undisturbed gas to the diameter of the hole. For a problem with very simple geometry, such as this, it would be expected that the transition between the two limits would be monotonic with the major change occurring between Knudsen numbers of 0·1 and 10. Liepmann (1961) discusses the effusion flow in depth and reports experiments that generally confirm these expectations. However, for more complex flows, the choice of Knudsen number may not be obvious and the transition between the two limits may not be monotonic.

The interaction of separate molecular streams and species in gas flows takes place through intermolecular collisions and, since these are absent in collisionless flows, a complex flow may be obtained by the superposition of simpler flows. Two consequences of this may be illustrated by further reference to the effusion problem. First, consider the case in which there is an equilibrium gas on both sides of the thin wall containing the small hole. As long as both mean free paths are extremely large in comparison with the dimensions of the hole, effusion will occur in both directions with each stream of gas being quite independent of the other. If we consider two vessels joined by one or more free molecule orifices and maintained at different temperatures, this leads to the phenomenon of *thermal transpiration* (see Exercise 5.1). Next, consider the free molecule effusion of a gas mixture rather than a simple gas. The components of the mixture effuse independently of one another and, since the mean molecular speed is inversely proportional to the square root of the molecular mass, there is a separation effect across the hole (see Exercise 5.2).

5.2. One-dimensional steady flows

A flow may be regarded as one-dimensional if it can be solved through the application of the molecular flux equations in one direction. The mass flux aspect of the molecular effusion problem qualifies as one-dimensional, even though the flowfield would be two or three-dimensional. This is the simplest class of flow and results from several additional examples are required for future reference.

Consider the *one-dimensional heat transfer* between two infinite plane parallel plates separated by a distance h that is very small in comparison with the mean free path λ of the gas between the plates. Assume diffuse reflection with complete thermal accommodation to the temperature T_U at the upper wall, and T_L at the lower wall. The monatomic gas of number density n between the plates then consists of an upward moving stream of density n_L and temperature T_L emitted from the lower plate, together with a downward moving stream of density n_U and temperature T_U emitted from the upper plate.

The flux of diffusely reflected molecules from a surface is identical with the flux of molecules from a fictitious equilibrium gas on the reverse side of the surface. The number density of the reflected molecules is just half that of the fictitious gas, since half of these molecules are moving in the direction normal to the surface. Therefore, from eqn (4.20),

$$n_U T_U^{\frac{1}{2}} = n_L T_L^{\frac{1}{2}}$$

and, since

$$n = n_U + n_L,$$

$$n_L = \frac{n T_U^{\frac{1}{2}}}{T_U^{\frac{1}{2}} + T_L^{\frac{1}{2}}}, \qquad n_U = \frac{n T_L^{\frac{1}{2}}}{T_U^{\frac{1}{2}} + T_L^{\frac{1}{2}}}. \tag{5.10}$$

The energy flux q_L from the lower plate is obtained by setting $\beta = (2RT_L)^{-\frac{1}{2}}$ and $\rho = 2n_L m$ in eqn (4.24) i.e.

$$q_L = mn_L\pi^{-\frac{1}{2}}(2RT_L)^{\frac{3}{2}}.$$

Similarly, the downward energy flux from the upper wall is

$$q_U = mn_U\pi^{-\frac{1}{2}}(2RT_U)^{\frac{3}{2}}.$$

The net upward heat flux is obtained by combining these equations with eqn (5.10), to give

$$q_f = -2^{\frac{3}{2}}\rho\pi^{-\frac{1}{2}}R^{\frac{3}{2}}T_U^{\frac{1}{2}}T_L^{\frac{1}{2}}(T_U^{\frac{1}{2}}-T_L^{\frac{1}{2}}). \tag{5.11}$$

The continuum heat transfer between the plates is

$$q_c = -K\frac{dT}{dy},$$

with the y axis normal to the lower plate. If we assume that $K = CT^B$ with C and B constant, this becomes

$$q_c = -\frac{C}{B+1}\frac{dT^{B+1}}{dy}.$$

The continuum energy equation in a static gas requires that q_c be a constant, so that

$$T^{B+1} = -\frac{(B+1)q_c}{C}y+\text{const.}$$

Since $T = T_L$ at $y = 0$, the constant of integration is T_L^{B+1}. Then, setting $y = h$, the continuum solution is

$$q_c = -\frac{C(T_U^{B+1}-T_L^{B+1})}{(B+1)h}. \tag{5.12}$$

For this flow, the free molecule and continuum solutions differ in their functional relationship as well as in the numerical constant. The free molecule heat transfer is proportional to the gas density and is independent of the plate spacing, whereas the continuum heat transfer is inversely proportional to the plate spacing and independent of the density. Therefore,

$$\frac{q_c}{q_f} \propto \frac{1}{\rho h} \propto \frac{\lambda}{h} \propto (Kn), \tag{5.13}$$

so that the ratio of the continuum to the free molecule heat transfer is proportional to the Knudsen number of the flow.

The free molecule heat transfer problem is readily extended to the *Couette flow* problem in which the lower plate remains stationary while the upper plate moves in the x direction (which lies in the plane of the plate) with velocity U. The number flux and the normal momentum flux at the plates is not affected by this velocity. The shear stress on the lower plate τ_L due to the particles from the upper plate having the mean velocity U in the x direction is obtained by setting $s = U\beta$, $\beta = (2RT_U)^{-\frac{1}{2}}$, $\rho = 2n_U m$, and $\theta = \pi/2$ in eqn (4.23). This gives

$$\tau_L = mn_U U(2RT_U/\pi)^{\frac{1}{2}}$$

or, using eqn (5.10),

$$\tau_L = \frac{\rho U}{(T_U^{\frac{1}{2}}+T_L^{\frac{1}{2}})}\left(\frac{2RT_U T_L}{\pi}\right)^{\frac{1}{2}}. \tag{5.14}$$

Note that the factor of 2 in the density again appears because the flux equations are applied to the fictitious gas on the opposite side of the upper plate and this has a number density of $2n_U$. The reflected molecules from the lower wall do not contribute to the shear stress so that τ_L is the net value. There will be an equal and opposite shear stress on the upper wall. The energy flux q_U from the upper wall may be obtained by making a similar substitution into eqn (4.24), to give

$$q_U = mn_U\pi^{-\frac{1}{2}}(2RT_U)^{\frac{1}{2}}(U^2/2+2RT_U). \tag{5.15}$$

In both the above cases, the gas between the two plates is macroscopically uniform. The combination of the two equilibrium distributions gives the following non-equilibrium distribution function for this gas;

$$\begin{aligned} f = \frac{1}{\pi^{\frac{3}{2}}n}(n_U\beta_U^3 \exp[-\beta_U^2\{(u-U)^2+v^2+w^2\}]+ \\ +n_L\beta_L^3 \exp\{-\beta_L^2(u^2+v^2+w^2)\}). \end{aligned} \tag{5.16}$$

The appropriate moments of this distribution may be evaluated to show that the velocity in the x direction is

$$u_0 = (n_U/n)U, \tag{5.17}$$

and the temperature is

$$T = \{T_L^{\frac{1}{2}}(T_U+U^2/3R)+T_U^{\frac{1}{2}}T_L\}/(T_U^{\frac{1}{2}}+T_L^{\frac{1}{2}}). \tag{5.18}$$

There is, therefore, a discontinuity in temperature and velocity between the surface of the plates and the adjacent gas. Such discontinuities are generally present at surfaces in free molecule flows.

5.3. One-dimensional unsteady flows

Consider a semi-infinite stationary uniform gas on the negative side of the plane $x = 0$ and separated by a thin wall from a vacuum on the positive side of the plane. At time $t = 0$, the thin wall is removed and the gas expands freely into the vacuum. Given that the gas is monatomic and originally at a temperature T_1 with number density n_1, the problem is to determine the number density, velocity, and temperature of the gas as a function of x and t; for values of x and t that are very small in comparison with the mean free path and mean collision time, respectively, in the flow. This is called the *free expansion* problem and is probably the simplest one-dimensional unsteady collisionless flow.

The collisionless Boltzmann equation provides the most direct method of tackling this problem. The one-dimensional form of eqn (5.6) is

$$n\bar{Q}(x,t) = \frac{1}{t}\int Qn_i f_{u_i}\left(\frac{x-x'}{t}, x'\right) \mathrm{d}x',$$

where (5.19)

$$x' = x - ut,$$

and the integration is over the initial configuration of the gas. The distribution function is given by eqn (4.13) as a function of $\beta_1 = (2RT_1)^{-\frac{1}{2}}$ and, setting $n_i = n_1$ and $Q = 1$, the density is

$$n(x,t) = \frac{1}{t}\int_{-\infty}^{0} n_1(\beta_1/\pi^{\frac{1}{2}}) \exp(-\beta_1^2 u'^2)\, \mathrm{d}x'$$

$$= \frac{n_1}{\pi^{\frac{1}{2}}}\int_{\beta_1 x/t}^{\infty} \exp\left\{-\frac{\beta_1^2(x-x')^2}{t}\right\} d\left\{\frac{\beta_1(x-x')}{t}\right\}.$$

Therefore, using the standard integrals of Appendix B,

$$\frac{n}{n_1} = \frac{1}{2}\operatorname{erfc}\left(\frac{\beta_1 x}{t}\right). \tag{5.20}$$

Similarly, the stream velocity is obtained by setting $Q = u$, to give

$$nu_0 = \frac{n_1}{\pi^{\frac{1}{2}}}\frac{1}{2\beta_1}\exp\{-(\beta_1 x/t)^2\}$$

or

$$\beta_1 u_0 = \pi^{-\frac{1}{2}} \exp\{-(\beta_1 x/t)^2\}/\operatorname{erfc}(\beta_1 x/t). \tag{5.21}$$

Note that eqns (5.20) and (5.21) are functions of $\beta_1 x/t$ alone and, at $x = 0$, $n/n_1 = \frac{1}{2}$ and $\beta_1 u = \pi^{-\frac{1}{2}}$. The number flux across the plane $x = 0$ is therefore constant and equal to $n_1/(2\pi^{\frac{1}{2}}\beta_1)$. A comparison with eqn (5.7) shows that this is identical with the number flux in the steady effusion problem. The behaviour of n/n_1 and $\beta_1 u_0$ as functions of $\beta_1 x/t$ is shown in Fig. 5.1. The density profile is symmetric about the origin, with the perturbation from the

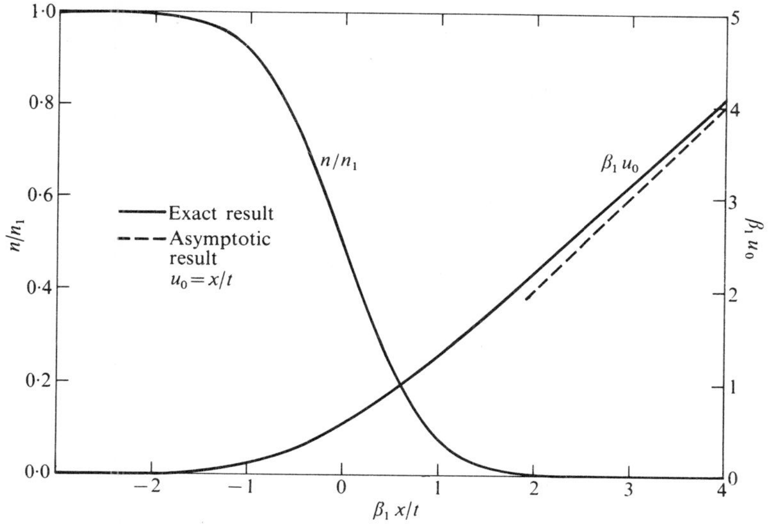

FIG. 5.1. Number density and stream velocity in a collisionless free expansion.

undisturbed densities falling as $\exp\{-(\beta_1 x/t)^2\}$. This latter result follows from the asymptotic expression for erfc(x) in eqn (C4). This result shows that, for large $\beta_1 x/t$, the velocity $\beta_1 u_0$ tends to $\beta_1 x/t$.

This type of analysis is not restricted to the determination of flow fields. For example, consider an extension of the above problem consisting of a specularly reflecting wall in the plane $x = x_w$. The problem is to determine the pressure p_w on the wall in the free molecule case for which $(Kn) = \lambda_1/x_w \to \infty$. The pressure on the wall is equal to twice the normal momentum flux at the location $x = x_w$, which may be obtained by setting $x = x_w$ and $Q = mu^2$ in eqn (5.19). Therefore

$$p_w = 2/t\int_{-\infty}^{0} n_1 mu^2 \frac{\beta_1}{\pi^{\frac{1}{2}}}\exp(-\beta_1^2 u^2)\,dx'$$

$$= \frac{2p_1\beta_1}{\pi^{\frac{1}{2}}}\int_{\beta_1 x_w/t}^{\infty}\left(\frac{x_w-x'}{t}\right)^2 \exp\left\{-\frac{\beta_1^2(x_w-x')^2}{t^2}\right\} d\left(\frac{x_w-x'}{t}\right),$$

and, again using the standard integrals,

$$\frac{p_w}{p_1} = 1-\operatorname{erf}\left(\frac{\beta_1 x_w}{t}\right)+\frac{2}{\pi^{\frac{1}{2}}}\left(\frac{\beta_1 x_w}{t}\right)\exp\left(-\frac{\beta_1^2 x_w^2}{t^2}\right). \qquad (5.22)$$

The analysis of the free expansion problem may also be extended to cover the expansion of a gas cloud of finite width. If this width is very small in comparison with the undisturbed mean free path, collisionless analysis is applicable at all locations and at all times. If the gas is initially uniformly distributed between the planes $x = -l$ and $x = +l$, the density and velocity are obtained similarly to eqns (5.20) and (5.21), but with the limits over x' from $-l$ to $+l$ rather than from $-\infty$ to 0. The results for the density and stream velocity are

$$\frac{n}{n_1} = \frac{1}{2}\left[\operatorname{erf}\left\{\frac{\beta_1(x+l)}{t}\right\} - \operatorname{erf}\left\{\frac{\beta_1(x-l)}{t}\right\}\right] \tag{5.23}$$

and

$$\beta_1 u_0 = \pi^{-\frac{1}{2}}\left[\exp\left\{-\frac{\beta_1^2(x-l)^2}{t^2}\right\} - \exp\left\{-\frac{\beta_1^2(x+l)^2}{t^2}\right\}\right] \div \div \left[\operatorname{erf}\left\{\frac{\beta_1(x+l)}{t}\right\} - \operatorname{erf}\left\{\frac{\beta_1(x-l)}{t}\right\}\right]. \tag{5.24}$$

The character of these solutions is shown in Fig. 5.2 for the number density and Fig. 5.3 for the flow velocity. The gas cloud is symmetrical about $x = 0$ at all times. For points within the initial extent of the cloud from $x = -l$ to $x = +l$, the density falls continuously, while the velocity increases to a maximum and then decreases. On the other hand, for points that are initially outside the cloud, the maximum velocity is at the leading edge of the expanding cloud, while the density increases to a maximum and then decreases. Narasimha (1962) deals also with the corresponding cylindrically and spherically symmetric flows.

The *one-dimensional piston problem* is another basic flow for which results are required for later reference. A specularly reflecting plane piston impul-

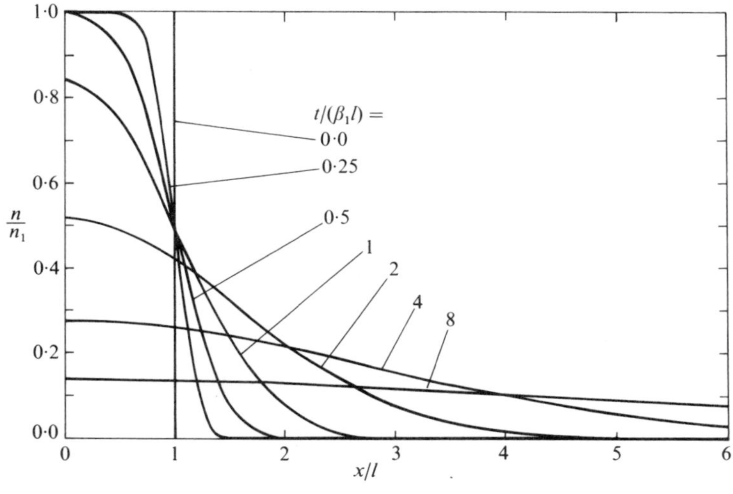

FIG. 5.2. Number density profiles during the one-dimensional expansion of a gas cloud.

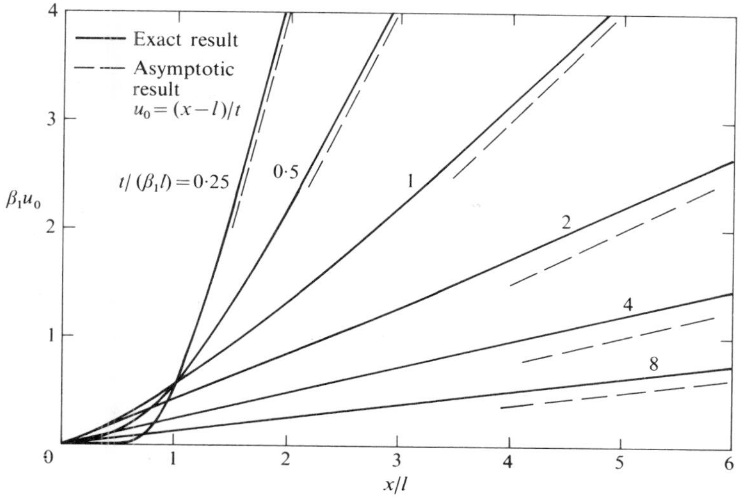

FIG. 5.3. Stream velocity profiles during the one-dimensional expansion of a gas cloud.

sively acquires a velocity $\pm U$ at time $t = 0$ and moves either into or away from an initially uniform stationary gas of density n_1 and temperature T_1. This problem may appear to be quite distinct from the free expansion problem. However, if the problem is viewed from a frame of reference moving with the piston face, the piston may be regarded as fixed plane at $x = 0$ with the gas on the positive side acquiring the velocity $-U$ towards the plane at time $t = 0$. Moreover, the plane $x = 0$ may be regarded as a surface of symmetry with a uniform gas of density n_1, temperature T_1, and velocity $+U$ as the initial state on the negative side. The collisionless Boltzmann equation (5.19) may now be applied to the problem.

The number density becomes

$$n = \frac{1}{t}\int_0^\infty n_1 \frac{\beta_1}{\pi^{\frac{1}{2}}} \exp\{-\beta_1^2(u+U)^2\}\,\mathrm{d}x' + \frac{1}{t}\int_{-\infty}^0 n_1 \frac{\beta_1}{\pi^{\frac{1}{2}}} \times$$
$$\times \exp\{-\beta_1^2(u-U)^2\}\,\mathrm{d}x',$$

the first term corresponding to the gas initially on the positive side of $x = 0$ and the second term to that initially on the negative side. In terms of the standard integrals, this becomes

$$n = \frac{n_1}{\pi^{\frac{1}{2}}}\left(\int_{-\beta_1(x/t+U)}^{\infty} \exp\left[-\beta_1^2\left\{\left(\frac{x-x'}{t}\right)+U\right\}^2\right] \mathrm{d}\left[-\beta_1\left\{\left(\frac{x-x'}{t}\right)+U\right\}\right] + \right.$$
$$\left. + \int_{\beta_1(x/t-U)}^{\infty} \exp\left[-\beta_1^2\left\{\left(\frac{x-x'}{t}\right)-U\right\}^2\right] \mathrm{d}\left[\beta_1\left\{\left(\frac{x-x'}{t}\right)-U\right\}\right]\right)$$

and may be evaluated to give

$$\frac{n}{n_1} = 1+\frac{1}{2}\left\{\operatorname{erf}\left(\frac{\beta_1 x}{t}+s\right)-\operatorname{erf}\left(\frac{\beta_1 x}{t}-s\right)\right\}, \tag{5.25}$$

where $s = U\beta$ is the speed ratio of the piston.

This solution for the density is illustrated in Fig. 5.4. The sign convention for U has been chosen such that a positive s corresponds to the piston moving into the gas and a negative s to the piston moving away from the gas. For positive values of s, the gas is doubled back with a speed $x/t = U$ relative to the piston or plane of symmetry, or $x/t = 2U$ in the frame of reference fixed in the stationary gas into which the piston moves. Superimposed on

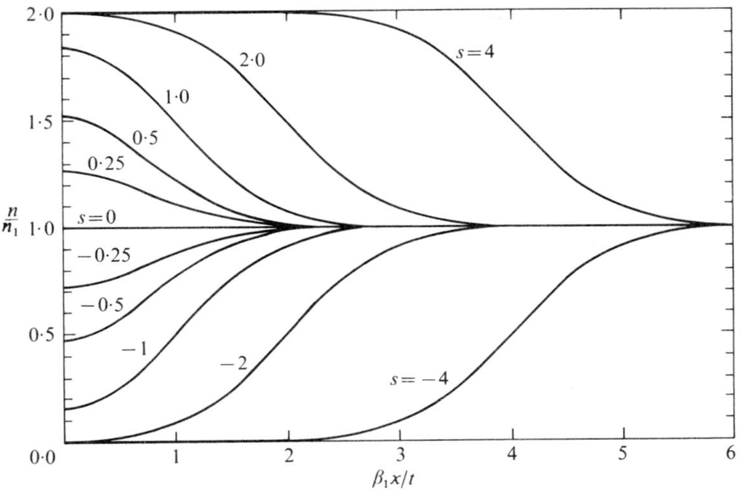

FIG. 5.4. Number density profiles for the one-dimensional collisionless piston problem.

this, there is a progressive spreading of the front due to the thermal motion of the molecules. The latter effect is dominant at small speed ratios, the number density at the piston face being

$$n/n_1 = 1+\operatorname{erf}(s). \tag{5.26}$$

For negative values of s, the profiles of n/n_1 lie between 1 and 0 and are symmetrical with those for positive values of s which lie between 1 and 2.

Solutions for the other flow field properties, such as the stream velocity, may readily be obtained by substituting the appropriate value of Q into eqn (5.19). The pressure on the piston face could be obtained by a similar calculation to that leading to eqn (5.22) for the free expansion impingement problem. It should be remembered that the pressure tensor is generally anisotropic in a free-molecule flow.

5.4. Free-molecule aerodynamics

This is concerned with the pressure, shear stress, and heat transfer at the surfaces of aerodynamic bodies in steady flight. The surface pressure and shear stress may be integrated over the complete body to determine the overall aerodynamic forces on the body. The properties of the flow field around the body may also be required in some applications. The Knudsen number is generally defined as the ratio of the mean free path in the undistrubed atmosphere to the characteristic dimension of the body, the free molecule solution being assumed to apply when this is sufficiently high. The reason for this assumption is that, under these circumstances, the molecules reflected from the body generally travel a very large distance before colliding with another molecule. There is then a negligible chance that the products of such collisions will collide with the body, so that the incident molecules may all be assumed to be from the equilibrium freestream. Note that the most significant collisions in the determination of the validity of the free molecule assumption are those between the reflected and the freestream molecules, and the mean free path for these may be much smaller than the freestream mean free path. This point is particularly important when the speed ratio, and therefore the Mach number, is large.

The surface properties follow from the application of the flux equations of § 4.2 to the incident and reflected molecules. The subscripts i and r will be used to denote the incident and reflected molecular streams; a flux being regarded as positive if it is directed towards the surface. In the absence of adsorption or emission effects at the surface, the incident number flux to a surface element must be balanced by the reflected number flux. Therefore,

$$N = N_{\mathrm{i}} + N_{\mathrm{r}} = 0; \tag{5.27}$$

$$p = p_{\mathrm{i}} + p_{\mathrm{r}}; \tag{5.28}$$

$$\tau = \tau_{\mathrm{i}} + \tau_{\mathrm{r}}; \tag{5.29}$$

and

$$q = q_{\mathrm{i}} + q_{\mathrm{r}}. \tag{5.30}$$

The results are obviously dependent on the nature of the gas–surface interaction and, in the absence of a general theory for such interaction, calculations will be made for a combination of the classical models of diffuse and specular reflection. From the discussion of these models in § 4.5 it can be seen that, for specular reflection,

$$p_{\mathrm{r}} = p_{\mathrm{i}} \quad \text{or} \quad p = 2p_{\mathrm{i}}, \tag{5.31}$$

$$\tau_{\mathrm{r}} = -\tau_{\mathrm{i}} \quad \text{or} \quad \tau = 0, \tag{5.32}$$

and

$$q_{\mathrm{r}} = q_{\mathrm{i}} \quad \text{or} \quad q = 0. \tag{5.33}$$

The only simplification to eqns (5.28) to (5.30) for diffuse reflection is

$$\tau_{\mathrm{r}} = 0 \quad \text{or} \quad \tau = \tau_{\mathrm{i}}. \tag{5.34}$$

The values of p_{i}, τ_{i}, and q_{i} for small surface element are given directly by eqns (4.21), (4.22), and (4.30), respectively. These equations provide p_{i}, τ_{i}, and q_{i} as functions of the angle θ between the unit normal vector and the direction of the freestream velocity U_{∞}, the freestream density $\rho_{\infty} = n_{\infty}m$, the parameter $\beta_{\infty} = (2RT_{\infty})^{-\frac{1}{2}}$ which is a function of the freestream static temperature T_{∞}, and the speed ratio of the freestream $s = U_{\infty}\beta_{\infty}$. Eqns (5.31) to (5.33) show that these results provide a complete solution for the specular reflection case.

In diffuse reflection, the molecules are brought to rest relative to the surface and are re-emitted with the equilibrium distribution corresponding to a temperature T_{r}. The quantities p_{r} and q_{r} are therefore given by the flux equations for stationary gas. From eqn (4.21)

$$p_{\mathrm{r}} = \frac{n_{\mathrm{r}}m}{4\beta_{\mathrm{r}}^2} \tag{5.35}$$

and, from eqn (4.28),

$$q_{\mathrm{r}} = \left(\frac{\gamma+1}{\gamma-1}\right)\frac{n_{\mathrm{r}}m}{8\pi^{\frac{1}{2}}\beta_{\mathrm{r}}^3}. \tag{5.36}$$

The number density n_{r} may be found from the condition, expressed in eqn (5.27), that the net number flux to the element is zero. Substituting from eqn (4.18) for the incident number flux and from eqn (4.20) for the reflected number flux into eqn. (5.27), we have

$$\frac{n_{\infty}}{2\pi^{\frac{1}{2}}\beta_{\infty}}[\exp(-s^2\cos^2\theta)+\pi^{\frac{1}{2}}s\cos\theta\{1+\operatorname{erf}(s\cos\theta)\}] = \frac{n_{\mathrm{r}}}{2\pi^{\frac{1}{2}}\beta_{\mathrm{r}}}.$$

Therefore,

$$n_{\mathrm{r}} = n_{\infty}\left(\frac{T_{\infty}}{T_{\mathrm{r}}}\right)^{\frac{1}{2}}[\exp(-s^2\cos^2\theta)+\pi^{\frac{1}{2}}s\cos\theta\{1+\operatorname{erf}(s\cos\theta)\}]. \tag{5.37}$$

If a fraction ε of the molecules is reflected specularly and the remaining fraction $1-\varepsilon$ is reflected diffusely, the above results may be combined to give the following results for the pressure, shear stress and heat transfer at the surface.

$$\frac{p}{p_\infty} = \frac{2\beta_\infty^2 p}{\rho_\infty} = \left\{\frac{(1+\varepsilon)}{\pi^{\frac{1}{2}}} s\cos\theta + \frac{(1-\varepsilon)}{2}\left(\frac{T_r}{T_\infty}\right)^{\frac{1}{2}}\right\}\exp(-s^2\cos^2\theta) + \\ + \left\{(1+\varepsilon)(\tfrac{1}{2}+s^2\cos^2\theta) + \frac{(1-\varepsilon)}{2}\left(\frac{T_r}{T_\infty}\right)^{\frac{1}{2}}\pi^{\frac{1}{2}} s\cos\theta\right\}\{1+\operatorname{erf}(s\cos\theta)\}. \tag{5.38}$$

$$\frac{\tau}{p_\infty} = \frac{2\beta_\infty^2\tau}{\rho_\infty} = \frac{(1-\varepsilon)s\sin\theta}{\pi^{\frac{1}{2}}}[\exp(-s^2\cos^2\theta)+\pi^{\frac{1}{2}}s\cos\theta\{1+\operatorname{erf}(s\cos\theta)\}]. \tag{5.39}$$

$$\frac{\beta_\infty^3 q}{\rho_\infty} = \frac{(1-\varepsilon)}{4\pi^{\frac{1}{2}}}\left(\left\{s^2+\frac{\gamma}{\gamma-1}-\frac{\gamma+1}{2(\gamma-1)}\frac{T_r}{T_\infty}\right\}[\exp(-s^2\cos^2\theta)+\pi^{\frac{1}{2}}s\cos\theta\,\times \right. \\ \left. \times\{1+\operatorname{erf}(s\cos\theta)\}]-\frac{1}{2}\exp(-s^2\cos^2\theta)\right). \tag{5.40}$$

Note that, with the sign convention that has been used in this chapter, the energy accommodation coefficient of eqn (4.62) becomes

$$a_c = \frac{q_i+q_r}{q_i+q_w} = \frac{q}{q_i+q_w}.$$

Therefore, T_r may be replaced by T_w in eqn (5.40) if the right hand side is multiplied by the accommodation coefficient a_c. For complete thermal accommodation, T_r may be replaced by T_w in all three equations. The temperature T_e of an insulated surface element at which $q = 0$ then follows from eqn (5.40) as,

$$\frac{T_e}{T_\infty} = \frac{2(\gamma-1)}{\gamma+1}\left(s^2+\frac{\gamma}{\gamma-1}-\tfrac{1}{2}\exp(-s^2\cos^2\theta)[\exp(-s^2\cos^2\theta)+\right. \\ \left. +\pi^{\frac{1}{2}}s\cos\theta\{1+\operatorname{erf}(s\cos\theta)\}]^{-1}\right). \tag{5.41}$$

The above equations may be applied directly to determine the aerodynamic forces on a thin flat plate at an incidence α to the stream. For the upper surface, $\theta = 90+\alpha$ while, for the lower surface, $\theta = 90-\alpha$. Eqn (5.38) may then be summed for the upper and lower surfaces to give the upward force per unit area normal to the surface of the plate as

$$F_N/p_\infty = \{2(1+\varepsilon)/\pi^{\frac{1}{2}}\}s\sin\alpha\exp(-s^2\sin^2\alpha)+(1-\varepsilon)(T_r/T_\infty)^{\frac{1}{2}}\pi^{\frac{1}{2}}s\sin\alpha+ \\ +(1+\varepsilon)(1+2s^2\sin^2\alpha)\operatorname{erf}(s\sin\alpha). \tag{5.42}$$

The rearward force per unit area parallel to the plane of the plate is given similarly by eqn (5.39) as

$$F_{\mathrm{P}}/p = \{2(1-\varepsilon)/\pi^{\frac{1}{2}}\}s\cos\alpha\{\exp(-s^2\sin^2\alpha)+\pi^{\frac{1}{2}}s\sin\alpha\,\mathrm{erf}(s\sin\alpha)\}.$$

The lift and drag forces on the plate may be denoted by L and D, respectively. The lift coefficient of a plate of area S is

$$C_{\mathrm{L}} = \frac{L}{\frac{1}{2}\rho_\infty U^2 S} = \frac{(F_{\mathrm{N}}\cos\alpha - F_{\mathrm{P}}\sin\alpha)S}{\frac{1}{2}\rho_\infty U^2 S} = \frac{(F_{\mathrm{N}}/p_\infty)\cos\alpha-(F_{\mathrm{P}}/p_\infty)\sin\alpha}{s^2}.$$

Therefore, substituting for F_{N}/p_∞ and F_{P}/p_∞,

$$C_{\mathrm{L}} = \frac{4\varepsilon}{\pi^{\frac{1}{2}}s}\sin\alpha\cos\alpha\exp(-s^2\sin^2\alpha)+\frac{\cos\alpha}{s^2}\{1+\varepsilon(1+4s^2\sin^2\alpha)\}\times$$
$$\times\mathrm{erf}(s\sin\alpha)+\frac{(1-\varepsilon)}{s}\pi^{\frac{1}{2}}\sin\alpha\cos\alpha\left(\frac{T_{\mathrm{r}}}{T_\infty}\right)^{\frac{1}{2}}. \tag{5.43}$$

The drag coefficient is

$$C_{\mathrm{D}} = \frac{D}{\frac{1}{2}\rho_\infty U^2 S} = \frac{(F_{\mathrm{N}}/p_\infty)\sin\alpha+(F_{\mathrm{P}}/p_\infty)\cos\alpha}{s^2}$$

and, similarly,

$$C_{\mathrm{D}} = \frac{2\{1-\varepsilon\cos(2\alpha)\}}{\pi^{\frac{1}{2}}s}\exp(-s^2\sin^2\alpha)+\frac{\sin\alpha}{s^2}\times$$
$$\times[1+2s^2+\varepsilon\{1-2s^2\cos(2\alpha)\}]\,\mathrm{erf}(s\sin\alpha)+\frac{(1-\varepsilon)}{s}\times \tag{5.44}$$
$$\times\pi^{\frac{1}{2}}\sin^2\alpha\left(\frac{T_{\mathrm{r}}}{T_\infty}\right)^{\frac{1}{2}}.$$

The heat transfer to the surfaces of the plate is given by eqn (5.40), and the temperatures of the upper and lower surfaces of an insulated plate are given by eqn (5.41). The total heat input to both sides of a perfectly conducting flat plate is given from eqn (5.40) as

$$\frac{(1-\varepsilon)a_{\mathrm{c}}\rho_\infty}{2\pi^{\frac{1}{2}}\beta_\infty^3}\left[\left\{s^2+\frac{\gamma}{\gamma-1}-\frac{\gamma+1}{2(\gamma-1)}\frac{T_{\mathrm{W}}}{T_\infty}\right\}\times\right.$$
$$\left.\times\{\exp(-s^2\sin^2\alpha)+\pi^{\frac{1}{2}}s\sin\alpha\,\mathrm{erf}(s\sin\alpha)\}-\tfrac{1}{2}\exp(-s^2\sin^2\alpha)\right].$$

The equilibrium surface temperature of an isolated perfectly conducting flat plate not subject to thermal radiation is the value of T_W for which this expression is zero, i.e.

$$T_W = \frac{2(\gamma-1)}{\gamma+1}\left[s^2+\frac{\gamma}{\gamma-1}-\tfrac{1}{2}\exp(-s^2\sin^2\alpha)\{\exp(-s^2\sin^2\alpha)+ + \pi^{\frac{1}{2}}s\sin\alpha\,\mathrm{erf}(s\sin\alpha)\}^{-1}\right]T_\infty. \quad (5.45)$$

For bodies other than flat plates, eqns (5.38) to (5.40) must be integrated over the surface. A list of existing solutions is given in Schaaf and Chambre (1961), but one frequently encounters a case that has either not been calculated or not reported in the literature. The chances of obtaining an analytical solution depend on the geometrical complexity of the problem. The procedure will be illustrated through the calculation of the drag coefficient of a sphere. For polar coordinates, as shown in Fig. 5.5, the drag coefficient is

$$C_D = \frac{D}{\frac{1}{2}\rho_\infty U^2\pi r^2} = \int_0^\pi (p\cos\theta+\tau\sin\theta)2\pi r^2\sin\theta\,d\theta/(\tfrac{1}{2}\rho_\infty U^2\pi r^2).$$

Patterson (1971) presents a useful table of the definite integrals which arise from the substitution of eqns (5.38) and (5.39) into this expression. The final result for the drag coefficient is

$$C_D = \frac{2s^2+1}{\pi^{\frac{1}{2}}s^3}\exp(-s^2)+\frac{4s^4+4s^2-1}{2s^4}\mathrm{erf}(s)+\frac{2(1-\varepsilon)\pi^{\frac{1}{2}}}{3s}\left(\frac{T_W}{T_\infty}\right)^{\frac{1}{2}}. \quad (5.46)$$

A remarkable feature of eqn (5.46) is that the fraction of specular reflection ε disappears from all terms except that containing the surface temperature T_W. Therefore, for a cold sphere, the drag coefficient is the same for both specular and diffuse reflection, with the common value tending to two as the speed ratio becomes large. However, it is most important that this result should not be interpreted as meaning that the drag coefficient of a sphere is independent of the gas–surface interaction model. The diffuse elastic model was mentioned in § 4.5, but was not pursued further because it is not

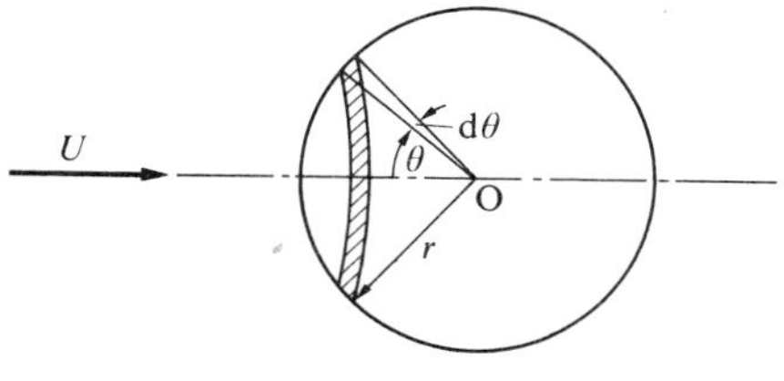

FIG. 5.5. Coordinate system for the analysis of sphere drag.

possible to obtain a closed analytical expression for the pressure due to the reflected molecules at an arbitrary speed ratio. However, this difficulty disappears for the high speed ratio case and the drag coefficient of a hypersonic sphere with diffuse elastic reflection is readily shown (see Exercise 5.10) to be three. It is therefore not inconveivable that the drag coefficient of, for example, a spherical satellite differs from the common diffuse and specular reflection value by as much as 50 per cent.

5.5. Thermophoresis

A thermal force acts on a small suspended particle in a gas with a temperature gradient, even though the temperature of the particle may be uniform and equal to the local gas temperature. The force is essentially due to the asymmetry of the Chapman–Enskog distribution function and the phenomenon is called thermophoresis. At sufficiently low densities or for sufficiently small particles, intermolecular collisions may be neglected and the magnitude of the effect is predicted by a free molecule analysis. In addition to the non-equilibrium reference gas, this example differs from the earlier collisionless flows in that the corresponding continuum flow is not well defined.

Consider a spherical particle of radius r and, as shown in Fig. 5.6, take the origin at the centre of the sphere and choose the x axis in the direction of the

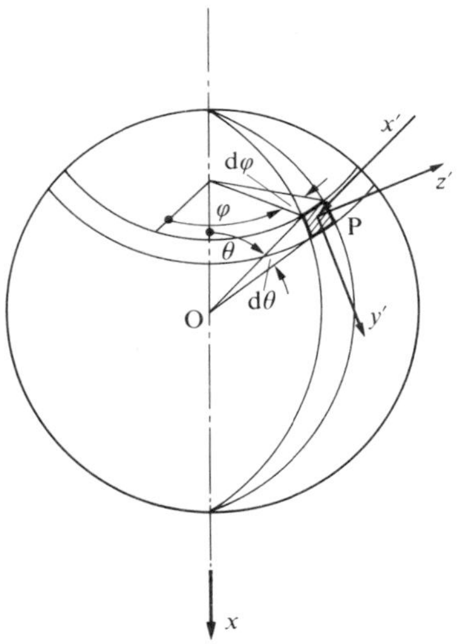

FIG. 5.6. Coordinate system for the analysis of the thermophoretic force on a sphere.

temperature gradient. The first approximation to the Chapman–Enskog distribution is then given by eqn (4.60) as

$$f = f_0\left\{1-\frac{4}{5}\frac{\beta^2 K}{\rho RT}\left(\beta^2 c^2-\frac{5}{2}\right)u\frac{\mathrm{d}T}{\mathrm{d}x}\right\}. \tag{5.47}$$

Now consider the small surface element at point P on the sphere defined by the polar angle θ and azimuth angle ϕ. Axes x', y', z' are then taken with x' normal to and directed towards the surface element and y' along the intersection of the planes defined by OPx and the surface element. Eqns (1.12) and (3.3) may then be applied in the x', y', z' frame, with f from eqn (5.47), to give the flux of the quantity Q to the surface element as

$$\frac{n\beta^3}{\pi^{\frac{3}{2}}}\int_{-\infty}^{\infty}\int_{-\infty}^{\infty}\int_{0}^{\infty} QU\exp\{-\beta^2(U^2+V^2+W^2)\}\times$$

$$\times\left[1-\frac{4}{5}\frac{\beta^2 K}{\rho RT}\left\{\beta^2(U^2+V^2+W^2)-\frac{5}{2}\right\}(U\cos\theta+V\sin\theta)\frac{\mathrm{d}T}{\mathrm{d}x}\right]\mathrm{d}U\,\mathrm{d}V\,\mathrm{d}W,$$

where U, V, and W are the velocity components in the x', y', and z' directions, respectively. By putting $Q = mU$ and $Q = mV$ we obtain, respectively, the pressure p_i and shear stress τ_i due to the impact of the incident molecules. The results are

$$p_{\mathrm{i}} = \frac{\rho}{4\beta^2}-\frac{2\beta K}{5\pi^{\frac{1}{2}}}\cos\theta\frac{\mathrm{d}T}{\mathrm{d}x}, \tag{5.48}$$

and

$$\tau_i = \frac{\beta K}{5\pi^{\frac{1}{2}}}\sin\theta\frac{\mathrm{d}T}{\mathrm{d}x}. \tag{5.49}$$

A term, such as the first one on the right hand side of eqn (5.48), that is uniform around the sphere obviously makes no contribution to the force on the sphere. A property of the Chapman–Enskog distribution is that the number flux is isotropic so that, if the surface temperature is uniform, the force exerted by diffusely reflected molecules is also uniform and makes no net contribution to the force. The pressure due to specularly reflected molecules is equal to p_{i}. The shear stress due to the diffusely reflected molecules is zero and that due to the specularly reflected molecules is equal to $-\tau_{\mathrm{i}}$. The net force is therefore given by

$$F = -\frac{\beta K}{5\pi^{\frac{1}{2}}}\frac{\mathrm{d}T}{\mathrm{d}x}\int_0^{\pi}\{2(1+\varepsilon)\cos^2\theta+(1-\varepsilon)\sin^2\theta\}2\pi r^2\sin\theta\,\mathrm{d}\theta,$$

where ε is the fraction of specularly reflected molecules. Therefore,

$$F = -\frac{16\pi^{\frac{1}{2}}}{15}\beta r^2 K\frac{\mathrm{d}T}{\mathrm{d}x}. \tag{5.50}$$

This result was first obtained by Waldmann (1959). Note that the disappearance of ε in the final result is due to the spherical geometry and would not occur for particles of a different shape.

5.6. The Rayleigh problem

Consider a semi-infinite volume of stationary uniform gas of density ρ_∞ and temperature T_∞ above the plane $y = 0$. This plane is a diffusely reflecting surface which, at time $t = 0$, acquires a temperature T_W and a velocity in the x direction of magnitude U_W. For times much less than the mean collision time and over distances much less than the mean free path, the resulting flow may be treated as collisionless. Gradients in the flow properties occur only in the y direction. However, in contrast with the strictly one-dimensional flows of § 5.3, there are finite flow velocities in both the x and y directions. The Rayleigh problem is useful as a test case when comparing the various methods for transition regime flows and our main purpose in treating it here is to establish the free molecule solution for later reference. A fuller discussion of the problem has been presented by Yang and Lees (1960).

The problem is best solved through the collisionless Boltzmann equation. Because finite velocities occur in two directions, we must refer back to eqn (5.6) and write

$$n\overline{Q}(y,t) = \frac{1}{t^3}\int\!\!\int\!\!\int Qn_\mathrm{i}f_\mathrm{i}\left(\frac{x-x'}{t},\frac{y-y'}{t},\frac{z-z'}{t},y'\right)\mathrm{d}x'\,\mathrm{d}y'\,\mathrm{d}z', \tag{5.51}$$

where $\boldsymbol{r}' = \boldsymbol{r}-\boldsymbol{c}t$, and the integration is over the gas at $t = 0$. The initial distribution function for $y' > 0$ follows from eqn (4.3) as

$$\frac{\beta_\infty^3}{\pi^{\frac{3}{2}}}\exp\{-\beta_\infty^2(u^2+v^2+w^2)\} = \frac{\beta_\infty^3}{\pi^{\frac{3}{2}}}\exp\left[-\beta_\infty^2\left\{\left(\frac{x-x'}{t}\right)^2+\left(\frac{y-y'}{t}\right)^2+\left(\frac{z-z'}{t}\right)^2\right\}\right].$$

The gas that is reflected from the surface when $t > 0$ may be represented at $t = 0$ as a gas at $y < 0$ with distribution function

$$\frac{\beta_\mathrm{W}^3}{\pi^{\frac{3}{2}}}\exp[-\beta^2\{(u-U_\mathrm{W})^2+v^2+w^2\}] = \frac{\beta_\mathrm{W}^3}{\pi^{\frac{3}{2}}}\exp\left[-\beta_\mathrm{W}^2\left\{\left(\frac{x-x'}{t}-U_\mathrm{W}\right)^2+\left(\frac{y-y'}{t}\right)^2+\left(\frac{z-z'}{t}\right)^2\right\}\right].$$

The number density of the gas at $y > 0$ is $n_\infty = \rho_\infty/m$, while that of the gas at $y < 0$ is n_{W}. The number flux in the y direction must be the same for each gas so that, from eqn (4.20),

$$n_{\mathrm{W}} = (\beta_{\mathrm{W}}/\beta_\infty) n_\infty .$$

Substituting these results into eqn (5.51), we have

$$\frac{n}{n_\infty}\bar{Q} = \frac{\beta_\infty^3}{\pi^{\frac{3}{2}} t^3}\int_{-\infty}^{\infty}\int_{0}^{\infty}\int_{-\infty}^{\infty} Q \exp\left[-\beta_\infty^2\left\{\left(\frac{x-x'}{t}\right)^2 + \left(\frac{y-y'}{t}\right)^2 + \left(\frac{z-z'}{t}\right)^2\right\}\right] \mathrm{d}x'\,\mathrm{d}y'\,\mathrm{d}z' + \frac{\beta_{\mathrm{W}}}{\beta_\infty}\frac{\beta_{\mathrm{W}}^3}{\pi^{\frac{3}{2}} t^3}\int_{-\infty}^{\infty}\int_{0}^{\infty}\int_{-\infty}^{\infty} \times$$
$$\times Q \exp\left[-\beta_{\mathrm{W}}^2\left\{\left(\frac{x-x'}{t} - U_{\mathrm{W}}\right)^2 + \left(\frac{y-y'}{t}\right)^2 + \left(\frac{z-z'}{t}\right)^2\right\}\right] \times$$
$$\times\, \mathrm{d}x'\,\mathrm{d}y'\,\mathrm{d}z'.$$

Solutions for the number density, stream velocity in the x direction, stream velocity in the y direction, and translational temperature are obtained by setting $Q = 1$, u, v, and $\frac{1}{3}c'^2/R$, respectively. The final results are

$$\frac{n}{n_\infty} = \frac{1}{2}\left[\left\{1 + \operatorname{erf}\left(\frac{\beta_\infty y}{t}\right)\right\} + \left(\frac{T_\infty}{T_{\mathrm{W}}}\right)^{\frac{1}{2}} \operatorname{erfc}\left(\frac{\beta_{\mathrm{W}} y}{t}\right)\right], \tag{5.52}$$

$$u_0 = \tfrac{1}{2} U_{\mathrm{W}}\left(\frac{T_\infty}{T_{\mathrm{W}}}\right)^{\frac{1}{2}} \operatorname{erfc}\left(\frac{\beta_{\mathrm{W}} y}{t}\right)\bigg/\left(\frac{n}{n_\infty}\right), \tag{5.53}$$

$$v_0 = \frac{1}{2\pi^{\frac{1}{2}}\beta_\infty}\left\{\exp\left(-\frac{\beta_{\mathrm{W}}^2 y^2}{t^2}\right) - \exp\left(-\frac{\beta_\infty^2 y^2}{t^2}\right)\right\}\bigg/\left(\frac{n}{n_\infty}\right), \tag{5.54}$$

and

$$T = T_\infty\left[1 + \tfrac{1}{2}\operatorname{erfc}\left(\frac{\beta_{\mathrm{W}} y}{t}\right)\left(\frac{T_\infty}{T_{\mathrm{W}}}\right)^{\frac{1}{2}}\left(\frac{T_{\mathrm{W}}}{T_\infty} - 1\right)\bigg/\left(\frac{n}{n_\infty}\right)\right] + \frac{y}{t}\frac{v_0}{3R} + \frac{U_{\mathrm{W}} u_0}{3R} - \frac{1}{3R}(u_0^2 + v_0^2). \tag{5.55}$$

The surface properties do not vary with time and may be obtained from the steady two-dimensional flow eqns (5.38) to (5.40) simply by putting $s = U_{\mathrm{W}}\beta_\infty$, $\theta = \pi/2$, $T_{\mathrm{r}} = T_{\mathrm{W}}$, and $\varepsilon = 0$. Therefore

$$p = \frac{\rho_\infty}{4\beta_\infty^2}\left\{1 + \left(\frac{T_{\mathrm{W}}}{T_\infty}\right)^{\frac{1}{2}}\right\}, \tag{5.56}$$

$$\tau = \frac{\rho_\infty U_{\mathrm{W}}}{2\pi^{\frac{1}{2}}\beta_\infty}, \tag{5.57}$$

and

$$q = \frac{\rho_\infty}{4\pi^{\frac{1}{2}}\beta_\infty^3}\left\{U_W^2\beta_\infty^2 + \frac{(\gamma+1)}{2(\gamma-1)}\left(1-\frac{T_W}{T_\infty}\right)\right\}. \tag{5.58}$$

The terms containing T_W represent the contribution of the reflected molecules and the remaining terms represent the contribution of the incident molecules. Note that while the flow field properties are presented in a frame of reference with the undisturbed gas at rest, the surface properties have been presented in a frame of reference fixed to the plate.

5.7. Flows with multiple reflection

The analysis of collisionless flows becomes much more difficult if the molecules that are reflected from a surface may reimpinge on the surface instead of escaping completely. Flows involving multiple surface interactions have practical application to internal flows and also to external flows past bodies of complex geometry.

Consider the surface S exposed to a gas under free molecule conditions. The number of molecules incident per unit time on a surface element dS at location S directly from the gas may be denoted by $N_1(S)\,dS$. Now let $P(S', S)$ be the probability that a molecule reflected from an element dS' at location S' strikes location S. The number of molecules per second which strikes dS as their second collision with the surface is, therefore,

$$N_2(S)\,dS = \int_S P(S', S)N_1(S')\,dS'\,dS.$$

Similarly, the number striking dS as their third collision is

$$N_3(S)\,dS = \int_S P(S', S)N_2(S')\,dS'\,dS$$

Therefore, is $N(S)$ is the total number flux at location S,

$$\begin{aligned} N(S) &= N_1(S)+N_2(S)+N_3(S)+\ ... \\ &= N_1(S)+\int_S P(S', S)\{N_1(S')+N_2(S')+N_3(S')+\ ...\ \}\,dS', \end{aligned}$$

or

$$N(S) = N_1(S)+\int_S P(S', S)N(S)\,dS'. \tag{5.59}$$

This is a Fredholm integral equation of the second kind and its solution must form part of the analysis of collisionless flows involving multiple reflection.

As an example, consider the free molecular flow through a circular tube of length b and radius r. The centreline of the tube lies along the x axis and the origin is at the entrance of the tube. The molecules that pass through the tube will be made up of those that pass directly through without collision with the inside surface, plus a fraction of those emitted from each element of the inside surface. The total flux N_t through the tube may therefore be written

$$N_t = N_d + \int_0^b N(x)P_e(x)\,\mathrm{d}x, \tag{5.60}$$

where N_d is the direct number flux, P_e is the probability that the molecules reflected from the element of length $\mathrm{d}x$ at x pass through the tube, and $N(x)$ follows from the axially symmetric version of eqn (5.59), i.e.

$$N(x) = N_1(x) + \int_0^b P(x', x)N(x')\,\mathrm{d}x. \tag{5.61}$$

Solutions of this problem were first obtained by Clausing (1932) and have been discussed in detail by Patterson (1971). The functions N_d, $P_e(x)$, $N_1(x)$ and $P(x', x)$ may be found for this simple geometry by straightforward but tedious analysis. However, the solution of the integral equation for $N(x)$ must be carried out numerically. The final integration of the eqn (5.60) requires the assumption that $N(x)$ is a linear function of x, with the coefficients provided by the numerical solution of eqn (5.61).

The above problem is of comparatively simple geometry and the predominantly analytical approach through the integral equation becomes quite unworkable for more complex problems. On the other hand, this class of problem is ideally suited to a probabilistic numerical approach called the *test-particle Monte Carlo method* (Davis 1960). This requires access to a digital computer, with the major demand being on central processor time rather than on storage capacity. Many thousands of typical molecular trajectories are generated in the computer and these provide a sufficiently large sample to predict the behaviour of the real system. Since intermolecular collisions may be neglected, these trajectories are independent of one another and may be generated serially. The method will now be illustrated through its application to the circular tube flux problem to calculate N_t and N_d.

The flow chart of the simulation program is shown in Fig. 5.7. The additional notation introduced by this chart is;

$L = b/r$ is the length–radius ratio of the tube, N_1 is the total number of particles entering the tube, r_1 is the initial radius of the particle at $x = 0$, and l_1, m_1, and n_1 are the direction cosines of the initial trajectory with the x, y, and z axes, x_c is the x coordinate of the intersection point with the cylinder

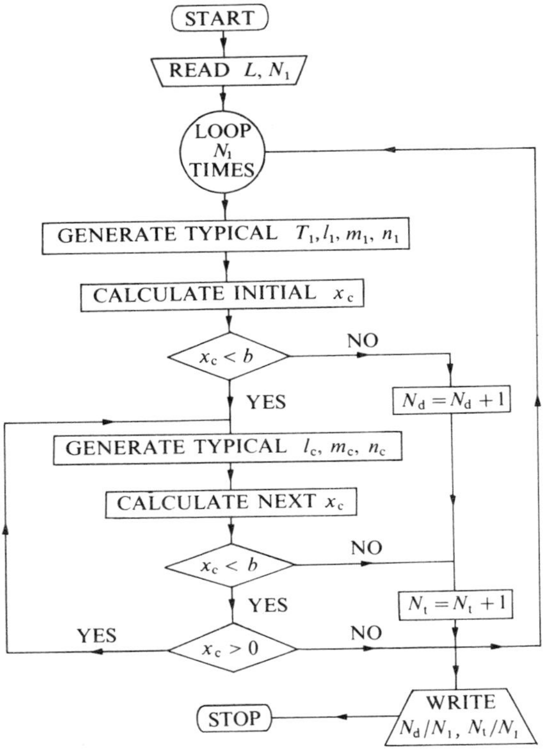

FIG. 5.7. Flow chart of the test particle Monte Carlo program for the cyindrical tube flux problem.

or its imaginary projections outside $x = 0$ to b, and l_c, m_c, and n_c are the direction cosines of the particle trajectory after diffuse reflection from the surface.

The distribution function of the initial radius is one of those dealt with in Appendix D, and the required expression follows directly from eqn (D6) as

$$r_1 = (R_f)^{\frac{1}{2}}r. \tag{5.62}$$

This chooses an initial radius at random from a uniform distribution over the entry plane. However, since the trajectories are independent, the scatter that is introduced into the initial conditions by this step may be avoided by setting a uniform distribution. The appropriate equation for the initial radius of the Nth trajectory is then

$$r_1 = \{N - 0{\cdot}5)/N_1\}^{\frac{1}{2}}r. \tag{5.63}$$

The number flux of molecules effusing across the entry plane of the tube is given by eqn (4.16) with $Q = 1$. The external gas is in equilibrium and the distribution function may be put into the polar coordinate form of eqn (4.4) with the polar direction along the axis of the tube. The u velocity component may then be written $c' \cos\theta$ and the number flux of molecules is

$$N_i = \frac{n\beta^3}{\pi^{\frac{3}{2}}} \int_0^{\infty} \int_0^{2\pi} \int_0^{\pi/2} c'^3 \exp(-\beta^2 c'^2) \sin\theta \cos\theta \, d\theta \, d\phi \, dc'. \tag{5.64}$$

The azimuth angle ϕ is uniformly distributed between 0 and 2π and a typical value is, from eqn (D5),

$$\phi = 2\pi R_f. \tag{5.65}$$

The distribution of the polar angle θ is best dealt with through the distribution of $\cos\theta$ since $\sin\theta \cos\theta \, d\theta = \cos\theta \, d(\cos\theta)$. The distribution function for $\cos\theta$ is therefore identical to that for the radius r which led, through eqn (D6), to eqn (5.62). Since the limits on $\cos\theta$ are from 0 to 1, we can write

$$\cos\theta = (R_f)^{\frac{1}{2}}. \tag{5.66}$$

The direction cosines l_1, m_1, and n_1 are then given by

$$\begin{aligned} l_1 &= \cos\theta, \\ m_1 &= \sin\theta \cos\phi, \\ \text{and} \quad n_1 &= \sin\theta \sin\phi. \end{aligned} \tag{5.67}$$

The x coordinate of the intersection of the particle trajectory with the cylinder, or its projections outside $x = 0$ to b, is readily obtained from elementary three-dimensional coordinate geometry. A general theorem with wide application to simulation studies is that the intersection of the line

$$\begin{aligned} x &= x_i + l_1 s \\ y &= y_i + m_1 s \\ z &= z_i + n_1 s \end{aligned} \tag{5.68}$$

and the quadric surface

$$\begin{aligned} F(x, y, z) \equiv a_{11}x^2 + a_{22}y^2 + a_{33}z^2 + 2a_{23}yz + 2a_{31}zx + 2a_{12}xy + \\ + 2a_{14}x + 2a_{24}y + 2a_{34}z + a_{44} = 0 \end{aligned} \tag{5.69}$$

is given by the roots of

$$A_1 s^2 + 2A_2 s + A_3 = 0, \tag{5.70}$$

where

$$A_1 = a_{11}l_1^2 + a_{22}m_1^2 + a_{33}n_1^2 + 2a_{23}m_1n_1 + 2a_{31}n_1l_1 + 2a_{12}l_1m_1,$$

$$A_2 = l_1(a_{11}x_i + a_{12}y_i + a_{13}z_i + a_{14}) + m_1(a_{21}x_i + a_{22}y_i + a_{23}z_i + a_{24}) + $$
$$+ n_1(a_{31}x_i + a_{32}y_i + a_{33}z_i + a_{34}),$$

and

$$A_3 = F(x_i, y_i, z_i).$$

Real roots of eqn (5.70) may be substituted into eqn (5.68) to determine the points of intersection. Since the present case is axially symmetric, we may choose $y_i = -r_1$ and $z_i = 0$. The coefficients a_{22} and a_{33} are equal to unity, $a_{44} = -r^2$, and all other coefficients are zero. Eqns (5.68) to (5.70) then give the following equation for the x coordinate of the intersection of the particle trajectory with the cylinder.

$$x_c = l_1[r_1m_1 + \{r^2(m_1^2 + n_1^2) - r_1^2n_1^2\}^{\frac{1}{2}}]/(m_1^2 + n_1^2). \tag{5.71}$$

Because of the axial symmetry, the y and z coordinates of the point of intersection may be set equal to $-r$ and 0, respectively. If the point of intersection lies within the cylinder, the assumption of the diffuse reflection model means that the selection of the direction cosines l_c, m_c, and n_c is similar to the above selection of m_1, l_1, and n_1, respectively. Note that the flux directions is now along the y, rather than the x, axis. The x coordinates of the subsequent point of intersection with the cylinder is, again using eqns (5.68) to (5.70),

$$x = x_c + 2l_crm_c/(m_c^2 + n_c^2).$$

This value of x becomes the next value of x_c.

A listing of a FORTRAN IV implementation of this program is provided as Appendix E. Note that this program contains only 32 executable statements. Results for the total flux ratio for a length–radius ratio of 1 are presented in Fig. 5.8 as a function of the sample size N. The value calculated by Clausing (1932) for this case is $N_t/N_1 = 0{\cdot}672$. The dashed lines represent this result $\pm 1/\sqrt{N}$, and it is clear that this provides a very good estimate of the expected scatter for a given sample size. The dots are based on the program in Appendix E with the initial radius set at random through eqn (5.62), while the crosses are based on the uniform spacing of eqn (5.63). The removal of the initial scatter does not appear to have reduced the final scatter in this example. A summation of all the results in Fig. 5.8 gives $N_t/N_1 = 0{\cdot}6713$ based on a sample of 201 720 trajectories. Similar agreement with Clausing's results was obtained for other values of the length–diameter ratio. A medium size computer such as the CDC6400 calculates trajectories at a rate of approximately one thousand per second.

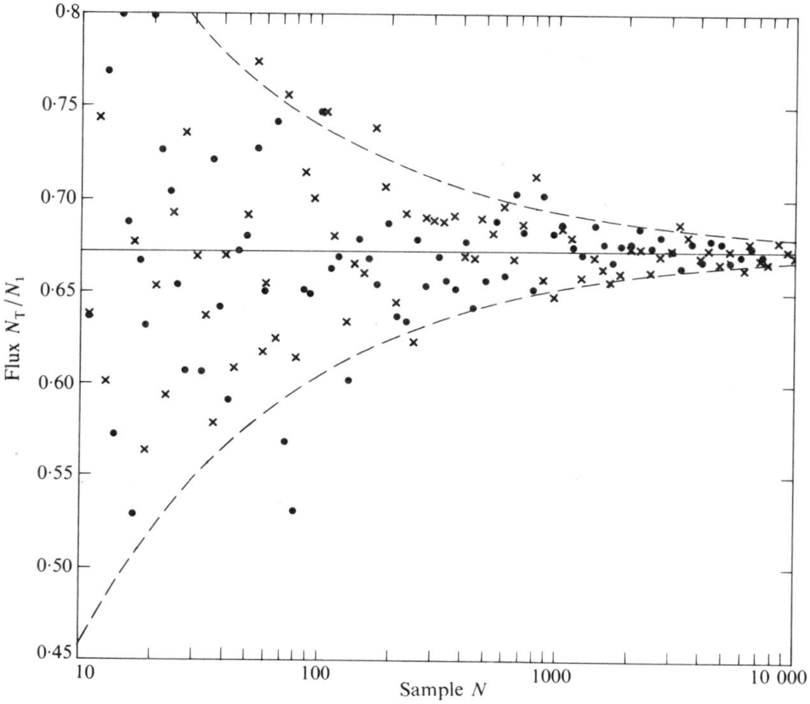

FIG. 5.8. Effect of sample size on the scatter in a typical Monte Carlo result ($L = 1$).

The major advantage of the Monte Carlo method is the ease with which it is able to deal with more complex flow geometries. Changes that would make the integral equation approach quite impractical can generally be made through alterations to several statements of the computer program. A number of configurations of engineering interest were dealt with by Davis (1960).

The above example required only the determination of the ratio of the exit to the entry number fluxes. When applying the Monte Carlo approach to problems such as the determination of a drag coefficient, it becomes necessary to relate the momentum transfer to the gas density. The simplest procedure for steady flows is to assume that the computed trajectories represent those occurring over unit time. The sum of the momentum transfer over all trajectories gives the momentum flux. The entry number flux may similarly be calculated and eqn (4.18) may then be used to obtain the corresponding number density. The test particle Monte Carlo method cannot be readily applied to unsteady flows and it is preferable to adopt a collisionless version of the direct simulation Monte Carlo method that is described in Chapter 7. In this, a large number of trajectories are computed simultaneously and there is a computational time parameter that can be directly related to real time.

Exercises

5.1. Consider two volumes of gas separated by a thin membrane containing holes. The typical dimensions of the holes are very small in comparison with the mean free paths in the gases. If the two volumes are maintained at temperatures T_A and T_B, but are otherwise undisturbed, show that the equilibrium pressures p_A and p_B in the two volumes are related by

$$\frac{p_A}{p_B} = \left(\frac{T_A}{T_B}\right)^{\frac{1}{2}}.$$

This phenomenon is called thermal transpiration.

5.2. Consider a binary gas mixture that is separated from a vacuum by a membrane of the type described in Exercise 5.1. The mass flux across the membane is sufficiently small for the gas to be regarded as being in equilibrium. The gas that effuses across the membrane is collected; the ratio of the number density of species 1 molecules to species 2 molecules being n_1'/n_2'. Show that this ratio is related to the original number density ratio by

$$\frac{n_1'}{n_2'} = \frac{n_1}{n_2}\left(\frac{m_2}{m_1}\right)^{\frac{1}{2}},$$

where m_1 and m_2 are the molecular masses.

5.3. A gas of density n_0 is effusing steadily through a hole of area S. Polar coordinates may be chosen with the origin at the hole and with the polar axis normal to the plane of the hole and in the direction of the flow. Use the result of Exercise 4.5 to show that the number density at large distance ($r^2 \gg S$) from the hole is

$$n = \frac{n_0 \cos\theta}{4\pi r^2} S.$$

Show further that, for a circular hole of radius r_0, the exact result for the number density along the axis is

$$n = \frac{n_0}{2}\left[1 - \left\{1 + \left(\frac{r_0}{r}\right)^2\right\}^{-\frac{1}{2}}\right].$$

Express this result as a power series in $(r_0/r)^2$ and examine the merging of the exact and approximate results.

5.4. Consider the one-dimensional heat transfer problem of § 5.2 with a constant gravitational acceleration g superimposed in the negative y direction. If $\bar{\rho}$ is the average gas density between the plates, show that the collisionless heat flux is

$$H_f = \frac{-2^{\frac{3}{2}}\pi^{-\frac{1}{2}}\bar{\rho}R^{\frac{1}{2}}gh(T_U - T_L)}{T_U^{\frac{1}{2}}[\exp(G_U^2)\{1-\operatorname{erf}(G_U)\}-1] + T_L^{\frac{1}{2}}[\exp(G_L^2)\{1+\operatorname{erf}(G_L)\}-1]},$$

where $G_U^2 = gh/(RT_U)$ and $G_L^2 = gh/(RT_L)$.

5.5. Simple Couette flow is the name given to the flow between a stationary surface in the plane $y = 0$ and a surface that is moving in the x direction with speed U in the plane

$y = h$. For incompressible flow with zero pressure gradient, the continuum solution for the velocity u is simply $u/U = y/h$. Show that the ratio of the collisionless shear stress for diffusely reflecting surfaces to the continuum shear stress is

$$\frac{\tau_f}{\tau_c} = \frac{8}{5\pi(Kn)}.$$

The Knudsen number (Kn) is defined as the ratio of the mean free path based on eqn (4.53) to the height h.

5.6. Consider the collisionless free expansion of a binary gas mixture. Show that the number density ratio is the semi-infinite gas case is

$$\left(\frac{n_2}{n_1}\right) \Bigg/ \left(\frac{n_2}{n_1}\right)_{\text{initial}} = \frac{\operatorname{erfc}\,[xm_2^{\frac{1}{2}}/\{(2kT)^{\frac{1}{2}}t\}]}{\operatorname{erfc}\,[xm_1^{\frac{1}{2}}/\{(2kT)^{\frac{1}{2}}t\}]},$$

where the subscripts 1 and 2 denote the two species. Plot the curve of n_2/n_1 versus at/x for an undisturbed gas composed of equal parts of argon and helium.

5.7. Show that the velocity field in the one-dimensional piston problem is given by

$$2\frac{n}{n_1}\beta u_0 = \pi^{-\frac{1}{2}}\left[\exp\left\{-\left(\frac{\beta x}{t}-s\right)^2\right\} - \exp\left\{-\left(\frac{\beta x}{t}+s\right)^2\right\}\right] - s\left\{\operatorname{erf}\left(\frac{\beta x}{t}+s\right) + \right.$$
$$\left. +\operatorname{erf}\left(\frac{\beta x}{t}-s\right)\right\},$$

where n/n_1 is the density ratio given by eqn (5.25).

5.8. Consider the two-dimensional free-molecule flow past a diffusely reflecting circular cylinder of diameter d and with its axis normal to the stream. Show that the number flux coefficient to the cylinder is

$$C_N = \frac{N}{nUd} = \frac{\pi^{\frac{1}{2}}\exp(-s^2/2)}{2s}\{(s^2+1)I_0(s^2/2)+s^2 I_1(s^2/2)\}$$

where N is the number of molecules striking the cylinder per unit length, and I_ν is the modified Bessel function of order ν.

The drag and heat flux coefficients may be written

$$C_D = \frac{D}{\frac{1}{2}\rho U^2 d} = \frac{\pi^{\frac{1}{2}}\exp(-s^2/2)}{s}\{(s^2+\tfrac{3}{2})I_0(s^2/2)+(s^2+\tfrac{1}{2})I_1(s^2/2)\} + \frac{\pi^{\frac{3}{2}}}{4s}\left(\frac{T_W}{T_\infty}\right)^{\frac{1}{2}}$$

and

$$C_H = \frac{H}{\frac{1}{2}\rho U^3 d} = \frac{1}{2}\left\{C_D - \frac{\pi^{\frac{3}{2}}}{4s}\left(\frac{T_W}{T_\infty}\right)^{\frac{1}{2}}\right\} - \frac{\gamma+1}{\gamma-1}\frac{C_N}{2s^2}\left(\frac{T_W}{T_\infty}-1\right),$$

where D and H are the drag and heat flux per unit length.

5.9. The 'cold surface hypersonic' limits for C_N, C_D, and C_H for the cylinder in the preceding exercise are 1, 2, and 1, respectively. Use the exact results to show that these are the zero order terms of a set of power series in $1/s^2$. Show that the first order results for a monatomic gas are

$$C_{\mathrm{N}} = 1+\frac{1}{4s^2},$$

$$C_{\mathrm{D}} = 2\left(1+\frac{3}{4s^2}\right),$$

and

$$C_{\mathrm{H}} = 1+\frac{11}{4s^2}.$$

Note that the corrections to the drag and heat transfer are, respectively, three and eleven times larger than that to the number flux.

5.10. The diffuse elastic model for the gas–surface interaction assumes that the molecules are reflected elastically, but in a completely random direction. Show tht this model leads to a value of three for the hypersonic limit of the drag coefficient of a cold sphere.

5.11. Consider the thermophoretic force on a thin conducting disc with its axis aligned with the temperature gradient in a gas. Show that the ratio of this force to that on a sphere of the same radius is equal to $3(1+\varepsilon)/4$.

6

GENERAL APPROACHES FOR TRANSITION REGIME FLOWS

6.1. Classification of methods

A TRANSITION regime flow is one in which the mean free path is neither very small nor very large in comparison with a characteristic dimension. The mathematical difficulties associated with the full Boltzmann equation generally preclude direct approaches that would lead to exact analytical solutions. This has meant that a large number of indirect approaches have been proposed and many of the resulting methods involve an unusual degree of approximation. It is therefore necessary to distinguish between a wide variety of competing methods. The answers to the following questions assist with this sorting process:

(*i*) Is the method analytical or numerical?

(*ii*) Is the method capable of handling all values of all parameters, or is it restricted to limiting cases in which one or more of the parameters is large or small?

(*iii*) What is the degree of complexity in terms of gas properties, boundary conditions, and number of dimensions that can be handled by the method?

(*iv*) Does the method require an initial approximation to the complete flow field and depend upon the convergence of a subsequent iteration process?

(*v*) Does the method depend on arbitrary assumptions about the functional form of certain parameters, or on arbitrary modifications to the basic equations?

Where any degree of approximation is involved, careful comparisons must be made with other theoretical results and with experiment. This is particularly important when the answer to questions (*v*) is affirmative. General conclusions should not be based on comparisons for a single case, or even for a single problem. Also, comparative tests should, as far as possible, be based on the full range of flow quantities over the complete flow field, rather than on the variation of a single integrated quantity with Knudsen number. The fact that solutions are available for the two limiting cases of collisionless and continuum flows means that superficially good results may be obtained from physically unreal methods that happen to provide a fortuitously good curve fit between these known limits.

In considering the first question, it must be kept in mind that some analytical methods provide 'solutions' that may require extensive numerical work in order to produce a result for a specific flow. In some cases, this numerical effort may exceed that which would be required for a direct numerical solution of the problem and the methods should then be regarded as providing alternative and perhaps less exact formulations rather than solutions. The direct numerical solutions are discussed in § 6.4.

Problems which involve large disturbances are invariably nonlinear and there is then no analytical method that does not incorporate assumptions or approximations. The small-perturbation approach may be based on the smallness of the Knudsen number or its reciprocal, in order to obtain approximate solutions for the near continuum or near collisionless regimes, respectively. Some methods may also require that the speed ratio be small or large, or that the flow disturbance itself be small. In particular, if the disturbance is sufficiently small that the distribution function is perturbed only slightly from the equilibrium or Maxwellian form, the small-perturbation approach leads to the linearized Boltzmann equation. The theory of this equation has been dealt with in detail by Cercignani (1969) and a more recent summary of progress may be found in Cercignani (1974). The small-perturbation solutions provide important reference values, especially when they are obtained without resort to any arbitrary assumptions about parameters or modifications to the equations. However, solutions are available only for a restricted set of problems with comparatively simple boundary conditions. While these may occasionally have direct engineering application, it is unlikely that they can be extended to cover new problems that are encountered in practice.

A number of methods depend on essentially arbitrary assumptions about the form of the distribution function. Most of these can be classed as moment methods and are discussed in § 6.2. The other area of major approximation is to alter the form of the Boltzmann equation itself. It is indicative of the difficulties that have been posed by transition regime flows that such a radical approach has been accepted and has led to useful results. The model equations are discussed in § 6.3.

There are, of course, a number of methods which do not fit readily into any general class. These are usually special treatments that are closely tailored to a particular flow geometry and offer little scope for extension to other flows.

6.2. Moment methods

This approach employs the moment equations that are obtained by first multiplying the Boltzmann equation by the molecular quantity Q, and then integrating it over velocity space. The moment equation for Q was derived as eqn (3.27) in the form

$$\frac{\partial}{\partial t}(n\overline{Q})+\nabla\cdot(n\overline{cQ})-nF\cdot\frac{\partial Q}{\partial c}=\Delta[Q]. \tag{6.1}$$

The macroscopic gas quantities were defined in § 1.2 in terms of the averages of the microscopic molecular quantities Q. The substitution of the various values of Q into eqn (6.1) leads to a series of equations in the macroscopic quantities, as described in § 3.3. However, the presence of $\overline{cQ}$ in the second term means that, as Q progresses to successively higher orders of c, each equation involves a moment of still higher order. We therefore have an infinite number of equations and moments. Eqn (3.3) enables a typical moment to be written

$$\overline{Q}=\int_{-\infty}^{\infty}Qf\,\mathrm{d}c$$

and the basis of the moment method is that f is assumed to conform to some expression containing a finite set of macroscopic quantities or moments. The higher order moment in one of the equations can then be written in terms of the lower moments, thus closing the series of equations and forming determinate set.

The Chapman–Enskog solution of the Boltzmann equation has already been discussed in §§ 3.5 and 4.4. It is the second order solution based on a series expansion of the distribution function f. This series may be written schematically as

$$f=f_0\{1+a_1(K_n)+a_2(K_n)^2+\dots\}, \tag{6.2}$$

where the coefficients a_n are functions of ρ, c_0, and T only. The Knudsen number is defined as the ratio of the scale length of the flow gradients to the mean free path. The first order† solution is the local equilibrium of Maxwellian distribution function f_0, with the gas being fully described by ρ, c_0 and T. The viscous stress tensor τ and the heat flux vector q vanish in an equilibrium gas and the conservation equations (3.32), (3.34), and (3.37) reduce to the Euler equations of inviscid fluid flow. The conservation equations are the five moment equations formed by setting Q equal to m, mc, and $\frac{1}{2}mc^2$ and, since these quantities are conserved in collisions, the collision term $\Delta[Q]$ vanishes. The second order Chapman–Enskog solution leads to the distribution function of eqn (4.58) which also involves only the moments ρ, c_0, and T. This distribution enables τ and q to be written as products of

† As far as the powers of (Kn) are concerned, this is the zero order solution. However, in § 3.5, the 'order' of the solution was based on the number of terms in the series and this convention is continued here.

the coefficients μ and K with the velocity and temperature gradients, respectively, thus reducing the conservation equations to the Navier–Stokes equations of continuum gas dynamics.

The above considerations show that, from the kinetic theory point of view, both the Euler and Navier–Stokes equations may be regarded as 'five moment' solutions of the Boltzmann equation, the former being valid for the $(Kn) \to 0$ limit and the latter for $(Kn) \ll 1$. In the context of continuum gas dynamics, the Euler equations describe reversible adiabatic (isentropic) flows, while the Navier–Stokes equations provide the well-established standard description of viscous flows. This means that first and second order Chapman–Enskog solutions merely reconcile the molecular and continuum approaches for small Knudsen-number flows. It is obvious that the expansion for f in eqn (6.2) must break down as the Knudsen number approaches unity, but there remains the possibility that the use of more than two terms may extend the range of validity of Chapman–Enskog theory to higher Knudsen numbers. The third order solution of eqn (6.2) leads to the Burnett equations. It has been suggested (Chapman and Cowling 1970) that these lead to a more refined description of the gas, as long as the Knudsen number is sufficiently low for the Chapman–Enskog theory to be valid. More specifically, Uhlenbeck (1974) suggests that the Burnett equations lead to a superior description of fast varying small disturbance flows. On the other hand, there is experimental evidence (Schaaf and Chambre 1961) that the inclusion of the higher order terms generally results in diminished agreement between theory and experiment as the upper Knudsen number limit of the Navier–Stokes equations is approached.

Grad (1949) put forward an alternative expansion for f as a series of Hermite tensor polynomials with the first-order solution again being the local Maxwellian. The third order solution leads to an expression for f as the local Maxwellian multiplied by an expression involving $\boldsymbol{\tau}$ and $\boldsymbol{q}$ as well as ρ, $\boldsymbol{c}_0$, and T. This means that the total number of moment equations required for a determinate set is equal to the number of dependent variables in the conservation equations. This number is readily seen to be thirteen (Exercise 3.2), thus leading to Grad's Thirteen Moment equations. While these represent a logical and important development in kinetic theory, the results from their application to specific problems in the transition regime have proved to be disappointing (Schaaf and Chambre 1961).

The failure of the systematic expansion methods to produce equations that clearly extend the upper Knudsen number limit of the Navier–Stokes equations indicates that more radical steps must be taken if workable approximations are to be found for the distribution function in the transition regime. For one-dimensional steady flow problems with two-point boundary conditions, the procedure has been to choose f as a combination of the two Maxwellian distributions appropriate to the boundaries. This may be a linear combination of the two full distributions, as in the Mott–Smith solution for shock wave structure (see § 8.1), or a two-stream combination, as in the Lees moment method (see § 8.2). The combination of Maxwellians introduces one or more additional dependent variables, thus requiring an equal number of moment equations to be added to the conservation equations. There are three conservation equations for a one-dimensional flow, so that the least number of moments is four. The collision term $\Delta[Q]$ must be evaluated for the non-conserved moment or moments and this generally brings the transport properties into the formulation. It is therefore possible to produce a closed form solution which spans the complete range of Knudsen number and which reduces to the collisionless solution as $(Kn) \to \infty$ and to the Navier–Stokes solution for $(Kn) \ll 1$.

Some of the moment methods lead to results that are in remarkably good agreement with experiment. However, these solutions are not unique because the method is based on several arbitrary choices. One of these is associated with the form of the assumed distribution function and the other with the choice of the non-conserved quantities for the additional moment equations. The best choice for one set of parameters is not necessarily the best for another set, even for the same problem. Also, the use of a more complicated distribution function, with a consequent increase in the number of moments, does not necessarily lead to more consistent results. The moment methods have not been applied successfully to two-dimensional problems such as the steady transition regime flow past a sphere. A major factor is that the form of the distribution function varies greatly from one location to another in such a flow. The difficulty in formulating a suitable expression for the distribution function is further complicated by the fact that it is three-dimensional rather than axially symmetric in velocity space. Finally, even if it were possible to find a workable approximation to the distribution function, the resulting equations would be more difficult to solve than the Navier–Stokes equations for the corresponding viscous continuum flow.

6.3. Model equations

The model equation approach involves an approximation to the Boltzmann equation itself. The major mathematical difficulties posed by the Boltzmann equation are associated with the collision term and it is this term that is modified. The best known model equation is called BGK equation after Bhatnagar, Gross, and Krook (1954). This may be written

$$\frac{\partial}{\partial t}(nf)+\boldsymbol{c}\cdot\frac{\partial}{\partial \boldsymbol{r}}(nf)+\boldsymbol{F}\cdot\frac{\partial}{\partial \boldsymbol{c}}(nf) = n\nu(f_0-f). \tag{6.3}$$

While the parameter ν is generally regarded as a collision frequency, it has a restricted functional dependence in that it is proportional to density and may depend on temperature, but is assumed to be independent of the molecular velocity $\boldsymbol{c}$. The appearance of the local Maxwellian f_0 means that the equation remains a nonlinear integro-differential equation. This is because f_0 is a function of the stream velocity $\boldsymbol{c}_0$ and the temperature T, and these must be obtained as integrals over f. These integrals over both f_0 and f must result in identical values of $\boldsymbol{c}_0$ and T in order for the collision term to vanish when moments are taken over the conserved quantities. This condition ensures that the BGK model is consistent with the conservation equations.

The BGK equation for stationary homogeneous gas without an external force field follows from eqn (6.3) as

$$\frac{\partial f}{\partial t} = \nu(f_0 - f)$$

and clearly has the correct solution $f = f_0$ at equilibrium. The Chapman–Enskog method may be applied to the BGK equation (Vincenti and Kruger 1965, p. 379). Although there are some differences in the numerical coefficients, the resulting transport properties convert the conservation equations to the Navier–Stokes equations. The BGK equation obviously provides the correct collisionless or free molecule solution, since the form of the collision term is then immaterial. An exact solution would therefore span the complete range of Knudsen number and would be close to the correct limits at the extremes of this range. The approximation to the collision term would, however, lead to an indeterminate error in the transition regime.

A comparison of eqns (6.3) and (3.20) shows that the depletion component of the collision term in the Boltzmann equation, $-\int_{-\infty}^{\infty}\int_0^{4\pi} n^2 f f_1 c_r \sigma \, d\Omega \, d\boldsymbol{c}_1$, is replaced in the BGK model by $-n\nu f$. The correct velocity dependent collision frequency $\nu(\boldsymbol{c})$ follows directly from the Boltzmann term as

$$\nu(\boldsymbol{c}) = n \int_{-\infty}^{\infty} \int_0^{4\pi} f_1 c_r \sigma \, d\Omega \, d\boldsymbol{c}_1 . \qquad (6.4)$$

This is independent of molecular velocity only for the physically unreal Maxwellian molecules, since eqn (2.33) shows that $\sigma \, d\Omega$ is then proportional to c_r^{-1}. The use of a velocity independent ν in the BGK model is generally an unjustified approximation, but at least the form of the depletion component is correct. The same cannot be said for the component of the collision term that represents the replenishing collisions. This is written $n\nu f_0$ and assumes that, irrespective of the form of f, the number of molecules scattered by collisions into class $\boldsymbol{c}$ is equal to the number that would be scattered out of this class in an equilibrium gas having the assumed velocity independent collision frequency.

The assumptions inherent in the BGK equation are least likely to have deleterious effects in problems that involve small perturbations and have their boundary conditions prescribed in terms of equilibrium distributions. The distribution function then departs only slightly from the Maxwellian form, irrespective of the value of the Knudsen number. This leads to linearized forms of the Boltzmann and BGK equations and, for these, the comparative mathematical simplicity of the model equation is particularly significant. Most of the results in Cercignani (1969) are for the BGK equation rather than for the full Boltzmann equation. For nonlinear problems, recourse must be had to moment methods or to numerical analysis.

6.4. Numerical methods

While numerical analysis is often required for the production of the final results from the moment and model equation methods, we are concerned in this section with methods that are not based on either an assumed form of the distribution function or on a modification of the Boltzmann equation. These methods may be divided into those which take the Boltzmann equation as their starting point and simulation methods that are based directly on the molecular description provided by kinetic theory.

The most direct numerical approach to the Boltzmann equation would be to construct a conventional finite difference formulation. The distribution function is the only dependent variable for a simple gas and this may be represented by its numerical value over a network of points in phase space. The first major problem arises from the sheer number of discrete points that would be required for this representation. For a problem with only one spatial dimension, the distribution function is axially symmetric in velocity space and a three dimensional array is required in phase space. There is the added complication that the velocity space domain is of infinite extent. Velocity space boundaries can be set such that only a negligible fraction of molecules lies outside them, but is difficult to predict the optimum placement of these. The distribution function becomes three-dimensional in two-dimensional flows, thus requiring a five-dimensional array of points in phase space. A three-dimensional array poses severe computational problems and these become almost insurmountable when the number of dimensions increases to five. The second major problem arises from the number of operations required to evaluate the collision term. The integrals are replaced by summations in the discrete approximation. At each point in phase space, a sum must be taken over all other velocity space points with each term in the sum representing, itself, a summation over the impact parameters of the collision. A direct finite difference approach would require a prohibitive amount of computer time and storage capacity for all but the simplest problems.

A finite difference approach has been applied to one-dimensional steady flow problems in a series of papers by Nordsieck, Hicks and Yen (for example:

Nordsieck and Hicks 1967, and Yen 1973). Their method uses a minimal number of cells in the axially symmetric velocity space and employs a Monte Carlo sampling technique to reduce the computing effort required by the summations associated with the collision term. Since the method is is applicable only to steady flows, an initial estimate must be made of the distribution function over the whole flow field, with the final solution being approached through an iterative process. The comparatively small number of cells leads to a rounding off of the velocities in collisions and a systematic error tends to build up from step to step in the iteration. A least squares procedure has been used to keep this in check and to ensure that the conservation laws are not violated. This correction procedure would be extremely difficult to apply to a two-dimensional flow, should it become computationally feasible to attempt such solutions. It is typical of finite difference methods that a particular flow configurations, or even particular sets of parameters for a given configuration, may require special modifications to the differencing scheme in order to maintain the stability of the calculation. Despite these difficulties, satisfactory results have been obtained for the structure of normal shock waves (Hicks, Yen, and Reilly 1972) and these will be discussed in § 8.1. The method has been applied only by its developers and its inherent difficulties and restrictions are such that it can hardly be regarded as a general purpose tool for solving problems in molecular gas dynamics.

The direct solution of the Boltzmann equation for the velocity distribution function does not constitute the sole means of obtaining solutions to transition regime flows. Advantage may be taken of the particulate nature of the gas to produce a direct simulation model on a digital computer. The gas is then represented by the positions and velocity components of a very large number of simulated molecules. While the distribution function may be established from the molecular information, the problem may be formulated without reference to it. Before taking this approach to its logical conclusion, we must mention a method that falls into an intermediate classification and is of some historical importance. This is best termed the *test particle Monte Carlo method* and is associated primarily with Haviland (1965).

The test particle method of collisionless flows has been presented in § 5.7 as the logical and simple method for dealing with flows involving multiple surface reflections. In the transition regime, it becomes necessary to compute typical intermolecular collisions in addition to the molecule–surface interactions. This can only be done if there is already a representation of the complete flow field. The distribution function is chosen for this representation and, as in the finite difference method, it is stored at a discrete number of points in phase space. The method therefore shares a major disadvantage of the finite difference method in that an initial estimate must be made of the distribution function over the whole flow field. A large number of test particle trajectories are computed with the assumed distribution serving as the 'target' gas for the computation of typical intermolecular collisions. A new target distribution is then constructed from the history of the test or incident molecules. This process is repeated and is assumed to have converged when there is no difference between the target and incident distributions.

The alternative to a test particle approach is to follow the trajectories of a very large number of simulated molecules simultaneously within a computer. This process commences from a specified initial state and then proceeds through a physically real unsteady process. Should the required flow be steady, it is obtained as the large time state of the unsteady flow. An initial estimate of the flow is not required and there is no iterative process. Alder and Wainwright (1958) were the first to adopt this direct simulation approach. This was in their *molecular dynamics* method which, apart from the setting of the initial configuration of the molecules, is completely deterministic. The trajectories of hard sphere molecules are computed and an intermolecular collision occurs whenever two trajectories converge to the molecular diameter. The impact parameters are provided by the orientation of the two trajectories. Application of the method has been severely restricted by the magnitude of the computing task. Each time a single molecular trajectory is considered, all other molecules must be examined as possible collision partners. This means that the computing task is proportional to the square of the number of molecules and, since the scatter due to the finite sample size decreases only as the square root of the sample, all but the most elementary problems are beyond the scope of the method.

The computing requirement becomes manageable if probabilistic rather than deterministic procedures are adopted for the computation of collisions. This leads to the direct simulation Monte Carlo method that was first applied to the homogeneous gas relaxation problem (Bird 1963) and is described in detail in the following chapter.

The above classification of numerical methods is not exclusive; alternative methods may be developed that are effectively new combinations of the various features. For example, Tuer and Springer (1973) have developed a method which is intermediate between the test particle and direct simulation Monte Carlo methods. Also, there are methods which do not fit the classifications of the previous sections and which hardly qualify as direct numerical methods. The discrete ordinate method that was first proposed by Broadwell (1964) provides a notable example. In this, a discrete set of points are chosen in velocity space, and the distribution function is described by its values at these points. The integro-differential Boltzmann equation reduces to a set of simultaneous, nonlinear, first order differential equation for the discrete values of the distribution function. Numerical methods are required for the solution of these equations for nontrivial flows.

7
THE DIRECT SIMULATION MONTE CARLO METHOD

7.1. General approach

THE direct simulation Monte Carlo method is a technique for the computer modelling of a real gas flow by some thousands of simulated molecules. The velocity components and position coordinates of the simulated molecules are stored in the computer and are modified with time as the molecules are concurrently followed through representative collisions and boundary interactions in simulated physical space. The flow chart of Fig. 7.1 is applicable to the computer programs for almost all applications.

Advantage may be taken of flow symmetries in order to reduce the number of position coordinates that need to be stored for each molecule. For example, in a one-dimensional flow with the flow gradients in the x direction, the x position coordinate is the only one that need be stored. All locations in the y, z plane are equally likely in such a flow. Similarly, only the radius need be stored in an axially or spherically symmetric flow, and only two position coordinates are required in a two-dimensional flow. The three velocity components in a Cartesian reference system are almost always stored for each molecule. This is because the collisions must be calculated as three-dimensional phenomena. Note that, if a radius is stored, a change in its direction must be accompanied by an appropriate rotation of the Cartesian reference frame for the velocity components.

The flow is always unsteady, but the boundary conditions may be such that a steady flow is obtained as the large time state of the unsteady flow. For example, if the objective is to study the steady flow past a cylinder, the initial state could be a uniform stream into which the cylinder is instantaneously inserted at zero time. Alternatively, the initial state could be a cylinder in a vacuum, with the stream commencing to enter from the upstream boundary at zero time. There is generally no choice in the specification of the initial state when the unsteady flow is to be studied. The essential and physically realistic unsteadiness of the flow means that it is not necessary to iterate from some initial approximation to a steady flow and there is consequently no question of convergence. It also means that the study of an unsteady flow presents no more difficulties than the study of a steady flow with the same number of dimensions in physical space.

The simulated region of physical space is divided into a network of cells. These cells may be either small regions with specified boundaries or simply

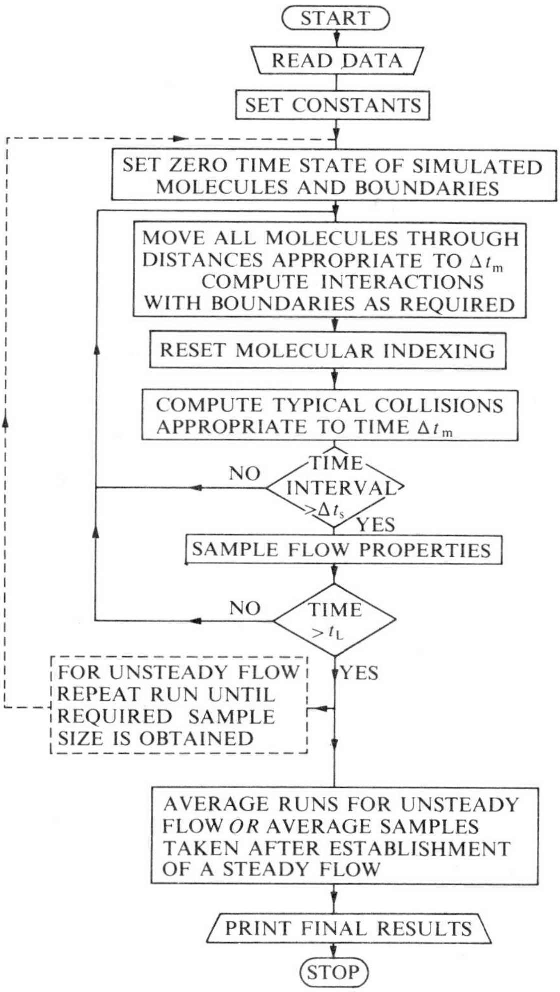

FIG. 7.1. Schematic flow chart for applications of the direct simulation Monte Carlo method.

an array of points. In the latter case, a molecule is said to be 'in a cell' when it is nearest the point which specifies that cell. The point reference scheme avoids the necessity of providing an analytical description of the cell boundaries and is, therefore, particularly useful for multi-dimensional problems with geometrically complex boundaries. The dimensions Δr of the cells must be such that the change in flow properties across each cell is small. Time is advanced in discrete steps of magnitude Δt_m, such that Δt_m is small compared with the mean collision time per molecule. Both Δr and Δt_m may vary with position and time.

The molecular motion and the intermolecular collisions are uncoupled over the small time interval Δt_m by the repetition of the following procedure:

(*i*) All the molecules are moved through distances appropriate to their velocity components and Δt_m. Appropriate action is taken if the molecule crosses boundaries representing solid surfaces, lines or surfaces of symmetry, or the outer boundary of the flow. New molecules are generated at boundaries across which there is an inward flux.

(*ii*) A representative set of collisions, appropriate to Δt_m, is computed among the molecules. The pre-collision velocity components of the molecules involved in the collisions are replaced by the post-collision values. Since the change in flow variables across a cell is small, the molecules in a cell at any instant may be regarded as a sample of the molecules at the location of the cell. This enables the relative positions of the molecules within the cell to be disregarded when choosing the collision pairs.

The computation of a representative set of collisions is of crucial importance in the simulation. This will now be discussed with specific reference to inverse power law molecules. The collision probability for a particular pair of molecules is shown by the discussion immediately preceding eqn (1.6) to be proportional to $\sigma_{\mathrm{T}} c_{\mathrm{r}}$ or, substituting for σ_{T} from eqn (2.27), $\pi W_{0,m}^2 (\kappa/m_{\mathrm{r}})^{2/(\eta-1)} c_{\mathrm{r}}^{(\eta-5)/(\eta-1)}$. If we use a constant $W_{0,m}$, and therefore a fixed deflection angle cut-off, the collision probability is proportional to the relative speed c_{r} raised to the power $(\eta-5)/(\eta-1)$ and the typical collision pairs must be selected on this basis. This is best done through the acceptance–rejection method (see Appendix D). Note that the exponent $(\eta-5)/(\eta-1)$ ranges from unity for hard sphere molecules to 0 for Maxwell molecules. The ability to select any two molecules from a particular cell is a prerequisite to the selection of collision pairs on the basis of $c_{\mathrm{r}}^{(\eta-5)/(\eta-1)}$. It is undesirable to keep a separate array for each cell for the storage of information on the molecules in the cell. Not only would this involve much transfer of information, but would be wasteful of storage since the number of molecules per cell changes in an unpredictable manner. An alternative would be to arrange the molecules in a single array in the order of their locations. This can sometimes be useful for one-dimensional unsteady problems since it permits a Lagrangian approach in which the cell location is set by the location of the molecules belonging to it. However, the continual re-ordering of the molecules requires a significant amount of computation time. This time has the particularly undesirable feature of being proportional to the square of the number of molecules in the computation. Derzko (1972) has put forward a more efficient scheme with computation time proportional to the product of the number of molecules and the number of cells. However, any procedure that is not linearly related to the number of molecules is to be avoided. The time penalty of such a procedure varies with the magnitude of the computation and precludes meaningful computation time comparisons with other methods. An efficient molecular referencing scheme, such that the computation time is directly proportional to the number of molecules, is described in § 7.3.

There remains the question of determining the appropriate number of collisions N_t over the time interval Δt_m. Eqn (1.7) shows that this number is $\frac{1}{2}N_m n \sigma_T c_r \Delta t_m$ or, again using eqn (2.27) for inverse power law molecules,

$$N_t = (N_m/2)\pi W_{0,m}^2(\kappa/m_r)^{2/(\eta-1)} n \overline{c_r^{(\eta-5)/(\eta-1)}} \Delta t_m, \tag{7.1}$$

where N_m is the number of molecules in the cell. One way of dealing with the problem would be to calculate either the number N_t of collisions per Δt_m or, preferably, sample the number from a normal distribution with N_t as mean. However, with the exception of the trivial case of Maxwell molecules, the computation of the mean value of $c_r^{(\eta-5)/(\eta-1)}$ over the sample of molecules in the cell is unduly time consuming. As with the molecular ordering, this procedure is especially undesirable because the required time is very nearly proportional to the square of the number of molecules in the cell. It is possible to completely avoid this step by advancing a time counter for each cell by the following amount at each collision;

$$\Delta t_c = (2/N_m)\{\pi W_{0,m}^2(\kappa/m_r)^{2/(\eta-1)} n c_r^{(\eta-5)/(\eta-1)}\}^{-1}. \tag{7.2}$$

Appendix F provides a formal proof that, over a large number of collisions, this procedure leads to the collision rate specified by eqn (7.1). It is worth quoting the special form of eqns (7.1) and (7.2) for hard sphere molecules with total cross section πd^2. These are

$$N_t = (N_m/2)\pi d^2 n \overline{c_r} \Delta t \tag{7.3}$$

and

$$\Delta t_c = (2/N_m)(\pi d^2 n c_r)^{-1}. \tag{7.4}$$

Sufficient collisions are calculated in each cell to keep the time counters concurrent with the overall time parameter. Note that Δt_c is essentially proportional to the inverse square of the number of molecules in the cell and, since $\overline{1/N^2}$ is significantly larger than $1/\overline{N}^2$ for small values of N, the statistical scatter may lead to an unrealistically low collision rate. It has been found that this effect is significant when N_m is less than about twenty. A solution is to use time or ensemble averaged values of N_m and n instead of the instantaneous values. Should this not be practicable, the value of Δt_c given by the inclusion of instantaneous values of n and N_m in eqn (7.4) may be reduced by an empirical correction factor. The factor $N_m^2/(N_m^2+12)$ has been found to give good results. Note also that we have not been concerned with the temporal distribution of the collisions within Δt_m. The implications of this point are discussed in the following section.

The cell network usually forms a convenient reference for the sampling of the macroscopic gas properties at intervals Δt_s. The number density n has already appeared in eqn (7.2), but this quantity is automatically sampled during the standard molecular indexing scheme. The velocity distribution may be sampled, but information is rarely required at this level. The fluxes of the various molecular quantities may be sampled at a boundary of the flow or at any surface immersed in the flow. In the case of solid boundaries, the incident and reflected fluxes may be sampled in the course of the simulation of the gas–surface interaction. A major advantage of the direct simulation method is that the boundary conditions are specified in terms of the probable behaviour of the individual molecules encountering the boundary, rather than by the form of the distribution function in the vicinity of the boundary.

Finite computer storage imposes limitations on both the total number of simulated molecules and the number of cells. The statistical scatter in the results is inversely proportional to the square root of the sample size. An ensemble average may be taken over repeated independent calculations of an unsteady flow in order to reduce the scatter to an acceptable level. If the flow becomes steady at large times, the required sample size may be built up by time averaging at sufficiently long intervals for the individual samples to be independent.

7.2. Relationship to the Boltzmann equation

The direct simulation method was developed through reference to the physics of the gas flow. This is in contrast with most numerical methods, such as finite difference methods, which are developed through reference to the mathematical description of the flow. The important question is the degree to which the simulation provides an accurate representation of the real flow. However, the strength of the mathematical tradition is such that more attention is usually given to the question of the relationship of the simulation solution to a hypothetical exact solution of the relevant equation, in this case the Boltzmann equation. The two questions would be identical if this equation provided an exact description of the real flow, but we saw in § 3.2 that there are limitations on the Boltzmann equation. These are its restriction to binary collisions, the treatment of these as instantaneous events, and the assumption of molecular chaos. These assumptions are readily seen to be implicit in the simulation procedures also. Therefore, as with Boltzmann equation, the validity of direct simulation method is dependent upon the gas being dilute.

The Boltzmann equation for inverse power law molecules may be written from eqns (3.20) and (2.27) as

$$\frac{\partial}{\partial t}(nf)+\boldsymbol{c}\cdot\frac{\partial}{\partial \boldsymbol{r}}(nf)+\boldsymbol{F}\cdot\frac{\partial}{\partial \boldsymbol{c}}(nf) = \pi W_{0,m}^{2}(\kappa/m_{\mathrm{r}})^{2/(\eta-1)}\int_{-\infty}^{\infty}\times$$
$$\times n^{2}(f'f'_{1}-ff_{1})c_{\mathrm{r}}^{(\eta-5)/(\eta-1)}\,\mathrm{d}\boldsymbol{c}_{1}. \tag{7.5}$$

We will now compare this equation with the simulation procedures and discuss the possible points of departure.

The first difference is that the dependent variable in the Boltzmann equation is the distribution function in phase space $\mathscr{F}$ ($\equiv nf$) which may describe an indefinitely large number of molecules. On the other hand, the simulation method deals with a discrete number of representative molecules. This leads to a significant degree of statistical fluctuation in the results. The magnitude of the expected fluctuations for a given sample size was discussed in § 1.2 and illustrated in Fig. 1.4. The general rule is that the standard deviation of the fluctuations is approximately equal to the square root of the sample size. It has been found that the scatter in the results is consistent with the total sample size irrespective of whether this is obtained from, for example, 100 samples with an average of 10 molecules per cell, or 10 samples with 100 molecules per cell. This provides reassurance that a reduction in the number of particles does not induce any systematic change or instability in the flow. Nevertheless, it is desirable to look more closely at the significance of the actual number of molecules involved in a particular flow.

In a typical application of the direct simulation method, a cell has dimension $\lambda/3$ and contains 30 molecules. For the typical molecular magnitudes of § 1.4, eqns (1.3) and (1.47) show that this corresponds to a ratio δ/d of 7, and coincides with the dilute gas limit that was employed in that section. Larger cells and/or a smaller number of molecules are frequently employed and it might appear that this would violate the dilute gas assumption. However, there are several independent reasons why this is not so.

(*i*) A molecular 'size' enters the simulation only through eqn (7.2) which effectively determines the collision frequency. As far as the molecular motion and boundary interactions are concerned, the molecules are assumed to have negligible dimensions. In fact, it is valid to consider each molecule as representing an arbitrarily large number of real molecules. In this case, the velocity components and position coordinates of the simulated molecules are the mean values of those of the real molecules. This approach is adopted when 'weighting factors' are applied to the molecules in order to equalize sample sizes in the simulation of gas mixtures with large concentration ratios, or flows which necessarily involve large variations in cell volume.

(*ii*) The cell dimensions are completely specified only in three-dimensional flows. When applied to a two-dimensional flow, the simulation may be regarded as applying to an arbitrarily thin 'slice' of the real flow. The thickness of the slice could be chosen such that the number of simulated molecules corresponds with the actual number. The cross-section of a one-dimensional flow is similarly adjustable.

The second argument leads to a physical interpretation of the statistical fluctuations. When the thickness of a two-dimensional flow or the cross-sectional area of a one-dimensional flow are chosen so that the number of real and simulated molecules are equal, the scatter is physically real and need not be regarded as an error. It would, or course, be desirable in most applications to increase the dimensions of the simulated region to such an extent that the scatter is negligible. However, we have seen that the standard deviation of the scatter falls only as the square root of the sample size and practical limitations on computer time mean that a certain level of scatter must be tolerated. This places a limit on the smallness of a disturbance that can be studied, since the 'signal' must be distinguished from the inevitable 'noise'.

A three-dimensional flow does not have any adjustable dimensions and the scatter has physical meaning only when the number of molecules in a cubic mean free path $n\lambda^3$ is the same in the real and simulated flows. For a equilibrium gas of molecules with fixed diameter $d = 3{\cdot}76 \times 10^{-10}$ m, eqns (1.45) and (1.47) give

$$n\lambda^3 = 5590(n_0/n)^2. \tag{7.6}$$

A typical simulation employs a cubic cell having a typical dimension $\lambda/3$ and containing 30 molecules; this corresponds to $n\lambda^3 = 810$. Therefore, while the number of molecules in a simulated flow will be much smaller than the real number at low densities, the two numbers may well correspond exactly in simulations above standard density.

Returning now to the Boltzmann equation itself, the left hand side of eqn (7.5) states that, in the absence of collisions, nf remains constant if one moves along with the molecules in phase space. The direct simulation Monte Carlo method traces the paths of the simulated molecules in phase space and this process is obviously consistent with the Boltzmann formulation. Therefore, any discrepancy between the simulation method and the Boltzmann equation would have to arise through the collision term on the right hand side.

The collision term is conveniently subdivided into the gain and loss terms, the latter representing the rate of scattering by collisions out of the phase space element $\mathrm{d}\boldsymbol{c}\,\mathrm{d}\boldsymbol{r}$ per unit volume of this element. This rate may be derived from the Monte Carlo procedure by the following argument.

The spatial cell in the simulated flow may be regarded as the element d$\boldsymbol{r}$ of physical space, and the set of N_m molecules in the cell defines the velocity distribution function f. The number of molecules of class $\boldsymbol{c}$ within the cell at time t is nf d$\boldsymbol{c}$ d$\boldsymbol{r}$, or $N_m f$ d$\boldsymbol{c}$. Now consider a collision between a molecule of this class with one of class $\boldsymbol{c}_1$. The probability of finding a molecule in class $\boldsymbol{c}_1$ is f d$\boldsymbol{c}_1$ and the time interval that would be added to a counter kept only for collisions of the class $\boldsymbol{c}$ molecule is given from eqn (7.2) as

$$\Delta t_c = \{\pi W_{0,m}^2 (\kappa/m_r)^{2/(\eta-1)} n c_r^{(\eta-5)/(\eta-1)}\}^{-1} \tag{7.7}$$

The factor $2/N_m$ has been omitted from eqn (7.2), the N_m because the time counter is for one rather than N_m molecules and the 2 because this molecule is just one of the collision pair. The collision rate for molecules of class $\boldsymbol{c}$ with those of class $\boldsymbol{c}_1$ follows as

$$N_m f \,\mathrm{d}\boldsymbol{c}\, f \,\mathrm{d}\boldsymbol{c}_1\, \pi W_{0,m}^2 (\kappa/m_r)^{2/(\eta-1)} n c_r^{(\eta-5)/(\eta-1)}$$

or

$$\pi W_{0,m}^2 (\kappa/m_r)^{2/(\eta-1)} n^2 f f_1 c_r^{(\eta-5)/(\eta-1)} \,\mathrm{d}\boldsymbol{c}_1\, \mathrm{d}\boldsymbol{c}\, \mathrm{d}\boldsymbol{r}.$$

The total rate of scattering out of class $\boldsymbol{c}$ per unit volume in phase space is obtained by integrating the class $\boldsymbol{c}_1$ over all velocity space and dividing by d$\boldsymbol{c}$ d$\boldsymbol{r}$. This gives

$$\pi W_{0,m}^2 (\kappa/m_r)^{2/(\eta-1)} \int_{-\infty}^{\infty} n^2 f f_1 c_r^{(\eta-5)/(\eta-1)} \,\mathrm{d}\boldsymbol{c}_1 \tag{7.8}$$

which agrees with the loss term on the right hand side of the Boltzmann equation (eqn 7.5). The gain term in this equation is derived by analogy with the loss term through the existence of inverse collisions for the simple monatomic gas models. The Monte Carlo method deals with a closed set of molecules from which both collision partners are selected, and the pre-collision velocity components are replaced by the post-collision values. This means that the correct choice of collision pairs, together with the appropriate random selection of impact parameters, automatically assures the correct gain term. This holds irrespective of the existence of inverse collisions and, on this point, the simulation method is less restrictive than the Boltzmann equation.

Since the procedures of the direct simulation method are consistent with the Boltzmann formulation, the results provide a solution of this equation as long as the computational approximations are kept within allowable bounds. These approximations are the use of a finite sample of molecules, the division of physical space into a finite number of cells, and the uncoupling of the molecular motion and the collisions over the small time interval Δt_m. In all applications of the method, decisions must be made about the number of molecules per cell, the cell size, and the time step. It would be expected that the best results would be obtained with the largest possible number of molecules, the cell size as small as possible in comparison with the mean free path, and Δt_m as small as possible in comparison with the mean collision time. In fact, the results are remarkably insensitive to these approximations and no deleterious effects are generally present if the number of molecules per cell drops as low as five or six, if the cell size approaches the mean free path, or if Δt_m approaches the mean collision time. However there are two additional points that should be made concerning these quantities.

The first concerns the use of eqn (7.2) to advance the cell time counter at each collision. Appendix F shows that this establishes the correct collision rate over a long period. However, if the number of molecules in a cell is very small, this may result in a physcially unreal correlation between the spacing of collisions and the probability of occurrence of the collisions. This is because the factor $2/N_m$ in eqn (7.2) is then comparatively large and, should an unlikely collision with a very small value of c_r occur, Δt_c may be much greater than Δt_m and no further collisions would occur for several time intervals. This correlation is effectively removed if the number of molecules N_m in a cell is large, so that the factor $2/N_m$ is small. Consider the simulation of a strong shock wave as an example. Collision pairs near the leading edge of the wave that include at least one molecule affected by the wave will generally have a much higher relative velocity and collision probability than pairs of undisturbed molecules. An unlikely collision between two undisturbed molecules would produce a large Δt_c and could lead to the above correlation. It has been found that there is a slight distortion of the upstream region of a very strong shock waves unless there are at least 20 molecules per cell.

The other point concerns the ratio of the typical cell dimension Δr to the time interval Δt_m. This ratio should not be large in comparison with the speed of propagation of acoustic disturbances. The reason for this additional condition is readily seen if we again consider the one-dimensional normal shock wave as an example. Assume that the shock is moving to the right and that, at a particular time, the leading edge of the wave has just penetrated the left-hand boundary of some cell. Since the actual locations of the molecules in a cell are disregarded when computing the collisions for a cell, one of the disturbed molecules at the far left of the cell may collide with one in the undisturbed gas near the right-hand boundary. This latter molecule will then be affected by the wave and could move into the next cell to the right when all the molecules are moved. It is therefore possible that a spurious signal could be propagated with speed $\Delta r/\Delta t_m$. However, this requires an improbable sequence of collisions affecting molecules near the boundaries of successive cells, and it has been found that this ratio must be several times the shock speed before the profile is affected. Moreover, the effective ratio may be reduced by placing a restriction on the spacing of molecules when choosing possible collision pairs.

The method has been applied to a wide variety of problems and there has never been any sign of instability in its operation. When the above requirements regarding sample size, cell dimensions, and time step are satisfied, the results are independent of these quantities and are repeatable to within the expected statistical scatter. Should these conditions not be satisfied, the deleterious effects consist of a reduction in the correct gradients of flow quantities, rather than as the growth of spurious gradients. Moreover, in order to see clear evidence of these effects one needs to look at a region of extreme non-equilibrium such as the leading edge of a strong shock wave.

7.3. Typical application

Many of the procedures that are required for the application of the direct simulation Monte Carlo method have already been introduced in § 5.7 in the context of the test particle Monte Carlo method for collisionless flows. Just as the details of that method were explained through its application to the cylindrical tube flux problem, the computational details of the direct simulation method are best illustrated by a typical application. The Rayleigh problem has been chosen as the subject of this application and the full FORTRAN program is listed as Appendix G.

We consider a gas of hard sphere molecules uniformly distributed between a diffusely reflecting wall in the plane $y = 0$ and a specularly reflecting wall in the plane $y = YM$. At zero time, the wall at $y = 0$ acquires the velocity UW in the x direction and its temperature jumps from the temperature of the undisturbed gas to TW times the temperature of this gas. The flow from $y = 0$ to YM is divided into NC cells of height CH, and each of these initially contains MC molecules. The time interval Δt_m is also read in as data and is denoted by DTM. The integers NIS and NST set the other times in the flow chart of Fig. 7.1 as multiples of Δt_m; the sampling time interval being $\Delta t_s = \text{NIS} * \text{DTM}$ and the limiting time $t_L = \Delta t_s * \text{NST}$.

The Rayleigh problem is essentially unsteady and the flow field is sampled at time intervals Δt_s, while the surface properties are sampled over these intervals. As indicated in Fig. 7.1, the runs are repeated in order to build up these samples; the final data variable NSEC effectively controls this process. NSEC is the amount of computer time available for the complete program and its use assumes that the elapsed computer time may be interrogated during the running of the program. This particular version of the program is written for the CDC 6000 or Cyber series and the subroutine CALL SECOND (A) returns the elapsed time in seconds as the variable A. The computer time is interrogated at the beginning and end of each run; a further run being made as long as the remaining time is at least fifty per cent greater than the time required for the preceding run. The coding for this is between the statement labelled 23 and that labelled 31.

Distances are normalized by the mean free path in the undisturbed gas λ_∞ and the value of this quantity is effectively regarded as unity within the program. The vertical extent of the flowfield YM is, therefore, the ratio of this quantity to λ_∞. Similarly, the most probable molecular speed is the undisturbed gas and its reciprocal β_∞ are regarded as having unit value, and this sets the normalization of the wall velocity UW. The undisturbed gas temperature and the molecular mass are also regarded as having unit value, so that the Boltzmann constant and the gas constant are both effectively equal to one half. The time DTM is normalized by $\beta_\infty \lambda_\infty$.

The necessary information on the simulated molecules is stored in the two-dimensional array P(L, M). The subscript L ranges from 1 to 4, with the velocity components in the x, y, and z directions being stored in L = 1 to 3, respectively. Since there are no gradients in the x and z directions, only the y position coordinate need be retained and this is stored in P(4, M). The subscript M may be regarded as the 'molecule number' and it is these numbers that are arranged in order of the cells in the cross-referencing array LCR, which also has the dimension M. The starting address of the molecules of cell N in this array is equal to IC(2, N) + 1, and the number of molecules in the cell is stored in IC(1, N). Three floating point numbers are required for each cell, and these are stored in the array C(3, L), where L is the maximum number of cells. The arrays SC (7, L, J) and SS (7, J) are used for sampling the flowfield and surface information, respectively. The remaining arrays are VRC(3) for the relative velocity components in a collision, and OP(10) for the temporary storage of output data.

The program follows the flow chart of Fig. 7.1. After the data input, values are set for the various constants required in the subsequent operations. It is convenient to keep the number density as a dimensional quantity based directly on the number of simulated molecules within a cell; this is stored in IC(1, N). Since the undisturbed number density n_∞ is set as FND = MC/CH, a density normalized by the undisturbed density is given by IC(1, N)/MC. The program is for a gas of hard sphere molecules, and eqn (4.38) shows that the mean free path in the undisturbed equilibrium gas is

$$\lambda_\infty = (2^{\frac{1}{2}}\sigma_T n_\infty)^{-1}.$$

Since the normalization procedure sets λ_∞ as unity, the collision cross-section σ_T is equal to $1/(2^{\frac{1}{2}} n_\infty)$ and is stored as CXS. The number of collisions per cell per Δt_m in the undisturbed gas, or ACU, follows similarly from eqn (4.37). The average relative speed of collision pairs VRM in the undisturbed gas is calculated from eqns (4.35) and (4.8).

The start of a run is at the statement labelled 20 and the first task is to set the initial or zero time configuration. The cell times C(2, N) are set equal to a random fraction of the mean time interval that would be added to them for a collision in the undisturbed gas. The use of this quantity instead of zero makes allowance for the fact that there must be an integer number of collisions over each DTM and that collisions are calculated until C(2, N) exceeds DTM. If this were not done, there would be an unrealistically high initial collision rate in a near free molecule flow. The initial approximation to the maximum relative speed in each cell is set, in C(3, N), to twice the average relative collision speed in the undisturbed gas. The other DO loops terminating at label 22 set the y coordinate and the u, v, and w velocity components of the MC molecules in each cell. The y coordinate is set at random within the cell, while the velocity components are generated through eqns (D9), (D10), and (D11).

The two DO loops over Δt_m and Δt_s are then set, the latter being a multiple of Δt_m through the integer NIS. The molecules are moved in the y direction over intervals appropriate to Δt_m within the loop terminating at label 25. First, the y coordinate is set as Y. If Y is greater than YM, the molecule has struck the specularly reflecting plane at y = YM. The routine for specular reflection from a surface is particularly simple in that the molecule is transferred to its image point with regard to the surface, while its velocity component normal to the surface is reversed. Similarly, if Y is less than 0, the molecule has struck the diffusely reflecting surface in the plane $y = 0$. In this case, the sampling variables for incident number, normal momentum, parallel momentum, and energy are advanced by the appropriate amounts. The surface information is sampled in a frame of reference in which the surface is stationary and the undisturbed gas moves with speed UW. The u and w components of the reflected molecular velocity may again be generated through eqns (D9) to (D11), although $1/\beta$ is now equal to VMW rather than unity. The distribution function for a velocity component normal to a diffusely reflecting surface follows from eqn (4.16) for the normal component of velocity in the molecular flux problem. This is

$$f_v = Cv \exp(-\beta^2 v^2),$$

where C is a constant. Now

$$f_v \, dv = Cv \exp(-\beta^2 v^2) \, dv = \tfrac{1}{2}C \exp(-\beta^2 v^2) \, d(v^2),$$

so that the distribution function for the square of the normal velocity component is

$$f_{v^2} = \tfrac{1}{2}C \exp(-\beta^2 v^2).$$

The constant C may be evaluated through the normalization condition of eqn (D2) to give

$$f_{\beta^2 v^2} = \exp(-\beta^2 v^2).$$

This distribution of v^2 is identical to the distribution of r^2 that leads to eqn (D11), so that

$$v = \{-\ln(R_f)\}^{\frac{1}{2}}/\beta. \tag{7.9}$$

The post-reflection value of P(2, N) is calculated from this relation, for which the appropriate value of $1/\beta$ is VMW. The amount of the time interval Δt_m that remains after the surface interaction is calculated as DTR, and the molecule is moved to the appropriate position above the surface. The sampling variables for the reflected properties are then advanced and the frame of reference for the u velocity component is changed to one in which the undisturbed gas is at rest.

The molecular indexing information in the arrays LCR(M) and IC(2, N) is not used prior to the collision routine and is most conveniently set between the molecular motion and collision segments of the program. The procedure for this necessarily samples the number of molecules per cell and, therefore, the number density. The number density is the only macroscopic quantity required by the collision procedures. The algorithm for molecular indexing is best deduced directly from the coding. The number of the cell in which a molecule lies is calculated twice for each molecule. This has been done here because of the geometric simplicity of the flow and the consequent brevity of this calculation. In general, it is preferable to calculate the cell number when moving the molecules and to store this in an integer array over the M molecules, say in IP(1, M). In that case, the LCR(M) array would preferably have been called IP(2, M) to avoid having two integer arrays over the M molecules, but note that M would not have the same meaning in each case.

The collisions appropriate to the time interval Δt_m are calculated within the loop over the number of cells NC and terminating at label 24. No collisions are calculated in a particular cell unless the time counter for the cell, C(2, N), is less than the overall flow time which is stored as the variable TIME. No collisions can be calculated unless there are at least two molecules in the cell. While this condition will almost always be met, failure to cover the contingency could result in an infinite loop. The numbers (i.e. addresses in the P(4, M) array) of the molecules in cell N are stored in the cross reference array LCR(M) for values of M between IC(2, N)+1 and IC(2, N)+IC(1, N). The statement labelled 32 selects a random value K for the address in the LCR array, and the resultant molecule number is L. A second molecule is then selected from the cell and given the number M. As long as M is not equal to L, these constitute a possible collision pair. The relative velocity between these molecules is then calculated and stored as VR and, since the program is for hard sphere molecules, the probability of collision is proportional to VR. The collision pair is therefore retained or rejected by the acceptance–rejection method of Appendix D, with the function f'_x of eqn

(D8) being equal to VR/C(3, N). If the collision pair is retained, the time interval for the collision is calculated from eqn (7.4) and is added to the cell time counter C(2, N). Since it is expected that the data for this program will always set at least 20 molecules per cell, instantaneous values of N_m and n may be used in eqn (7.4) without any correction factor for the effects of statistical scatter.

The collision mechanics of hard-sphere molecules was dealt with in § 2.4. The post collision velocities are most easily calculated through a direct application of the result that all directions are equally likely for the relative velocity after collision $\boldsymbol{c}_r^*$. The magnitude of this velocity was shown in eqn (2.8) to be unaffected by an elastic collision, so that it remains equal to VR in this program. An element of solid angle in polar coordinates, with θ as the polar angle and ϕ as the azimuth angle, is $\sin\theta \, d\theta \, d\phi$. Therefore, ϕ is uniformly distributed between 0 and 2π, while θ is between 0 and π with a distribution function $f_\theta = \sin\theta$. Now the fraction of angles between θ and $\theta + d\theta$ is

$$f_\theta \, d\theta = \sin\theta \, d\theta = -d(\cos\theta),$$

and this is also the fraction of molecules with cosines between $\cos\theta$ and $\cos\theta + d(\cos\theta)$. Therefore,

$$f_{\cos\theta} \, d(\cos\theta) = -d(\cos\theta) \tag{7.10}$$

so that $f_{\cos\theta}$ is a constant and $\cos\theta$ is uniformly distributed between -1 and $+1$. The collision routine follows immediately after the advancing of the collision number counter SC(7, N, J); the first statement being to set the working variable B as $\cos\theta$ by generating a number from the rectangular distribution between -1 and $+1$. The working variable A is then set as $\sin\theta$ and the three post-collision components of the relative velocity are obtained by multiplying VR by $\cos\theta$, $\sin\theta\cos\phi$, and $\sin\theta\sin\phi$, respectively. The post-collision velocity components follow from a direct application of eqns (2.1) and (2.5).

The earlier loop setting Δt_s as NIS times Δt_m also terminates at label 24 and the flowfield sampling routine follows as the three DO loops terminating at label 23. The outer loop is over the number of cells and sets the variable K equal to the number of molecules in the cell. The middle loop is over this number; the cross referencing arrays being used to select the molecules from the appropriate cell. Had the cell number been stored for each molecule, a simpler routine with a single DO loop over the molecules would have been possible. The quantities sampled are the sample size N, Σu, Σv, Σu^2, Σv^2, and Σw^2.

When all the runs have been completed, the sampled information is put into the required form and printed. The surface properties are based on the sampled flux of molecular number, momentum, and energy incident on and reflected from the surface. Since the multiplicative factor for the velocity components in the normalization is effectively unity, the required non-dimensional fluxes are obtained by dividing the sums of the appropriate quantities by the product of the number of runs, the sampling interval, and the undisturbed number density. The flow field properties are calculated from the sums of N, u, v, u^2, v^2, and w^2 for each cell at each sampling time. The normalized density is obtained by dividing N by the product of the number of runs and the number of molecules per cell in the undisturbed gas. The velocities in the x and y directions are simply $\bar{u} = \Sigma u/N$ and $\bar{v} = \Sigma v/N$, respectively. The temperature based on the u components of velocity is given from eqn (1.24) and Exercise 1.5 as

$$T_x = (m/k)\overline{u'^2} = (m/k)\{\overline{u^2} - \bar{u}^2\}$$

or, for the normalization used in this program,

$$T_x = 2(\Sigma u^2/N - \bar{u}^2).$$

Note that the result of Exercise 1.5 means that it is not necessary to sample the peculiar velocity components. The overall temperature is the mean of the three component temperatures.

Some typical results from the program are discussed in § 8.4. The above application is relatively simple and a number of additional direct simulation programs and routines are listed and/or described in subsequent chapters.

8

ONE-DIMENSIONAL FLOWS OF A SIMPLE MONATOMIC GAS

8.1. Shock wave structure

A SHOCK wave involves the transition from a uniform upstream flow to a uniform downstream flow. The flow is one-dimensional in that there are no flow gradients in directions parallel to the plane of the wave, and a frame of reference is usually chosen such that there is zero stream velocity in this plane. A shock Mach number $(Ma)_s$ may then be defined as the ratio of the speed of the wave, relative to the upstream gas, to the speed of sound in this gas. For the steady flow frame of reference in which the shock wave is stationary, the shock Mach number may be identified with the flow Mach number of the upstream gas. The pressure, density temperature, and steady stream velocity ratios across the wave are given by the Rankine–Hugoniot jump relations (Liepmann and Roshko 1957). These are based on the conservation of mass, momentum, and energy from the upstream to the downstream uniform states and are functions of the specific heat ratio and the shock Mach number only. The internal structure of the wave is, however, determined by viscous and heat conduction effects. The problem of the internal structure of shock waves has played an important role in the development of molecular gas dynamics. This is because it is a flow that involves a marked degree of thermal nonequilibrium for large values of $(Ma)_s$, it does not involve the uncertainties associated with solid boundaries, and has proved accessible to reliable experimentation.

The Navier–Stokes equations may be applied to the shock structure problem and yield an ordinary differential equation that is amenable to numerical solution (Gilbarg and Paolucci 1953). A closed form solution for a gas with a Prandtl number of 0·75 had been obtained earlier by Morduchow and Libby (1949). A shock wave thickness may be defined as the distance required to span the upstream to downstream density change by the maximum density gradient within the wave. Morduchow and Libby found that the ratio of this thickness to the average mean free path in the wave typically ranges from ten at $(Ma)_s = 1{\cdot}25$, five at $(Ma)_s = 1{\cdot}5$, three at $(Ma)_s = 2$, down to approximately two for shock Mach numbers above 3. These figures indicate that the Navier–Stokes theory is almost certain to be valid for shock Mach numbers less than 1·25, but is most unlikely to be valid for $(Ma)_s$ greater than two. Sherman (1955) and Talbot and Sherman

(1959) measured steady shock wave profiles in a low density wind tunnel at shock Mach numbers up to 1·8 and obtained excellent agreement between the experimental and Navier–Stokes profiles. The Burnett profile differs significantly from the Navier–Stokes profile for shock Mach numbers of the order of 1·8; these being among the first results to cast doubt on the validity of the Burnett formulation. The thirteen moment equations cannot be solved for the shock wave structure at shock Mach numbers above 1.65. Experimental results for very strong ($(Ma)_s \gg 2$) shock were obtained by Linzer and Hornig (1963), Camac (1965), Russell (1965), Schultz-Grunow and Frohn (1965), and Schmidt (1969). Schmidt's profiles were obtained from highly accurate electron beam measurements in a large low density shock tube. The experiments confirmed that the Navier–Stokes theory is inadequate when the shock Mach number is significantly greater than two and that the problem then falls within the transition regime. Three general approaches for the solution of transition regime flows were outlined in Chapter 6 and these have all been applied to the strong shock structure problem.

A moment method was applied by Mott-Smith (1951). He assumed that the distribution function within the wave may be represented as a linear combination of the equilibrium distribution functions that apply in the uniform upstream and downstream flows. This is sometimes called a *bimodal distribution* and may be written

$$nf = N_1 f_1 + N_2 f_2, \tag{8.1}$$

where the subscripts 1 and 2 apply to the upstream and downstream states, respectively. Taking the x axis normal to the wave, eqn (4.1) for the equilibrium or Maxwellian distribution gives

$$f_1 = (\beta_1^3/\pi^{\frac{3}{2}}) \exp[-\beta_1^2\{(u-u_{01})^2+v^2+w^2\}]$$

and

$$f_2 = (\beta_2^3/\pi^{\frac{3}{2}}) \exp[-\beta_2^2\{(u-u_{02})^2+v^2+w^2\}].$$

The weighting factors N_1 and N_2 must be such that $N_1 = n_1$, $N_2 = 0$ upstream of the wave and $N_1 = 0$, $N_2 = n_2$ downstream of the wave. Eqn (8.1) may be integrated over all velocity space and eqn (3.2) then requires

$$n = N_1 + N_2. \tag{8.2}$$

The application of the three conservation equations between the upstream and downstream states leads to the Rankine–Hugoniot solution for n_2, u_{02}, and β_2 as function of n_1, u_{01}, and β_1. The evaluation of all the flow quantities within the wave as the appropriate moments of the bimodal distribution function of eqn (8.1) requires a solution for n, N_1, and N_2 as functions of position within the wave. One relationship between these quantities has already been provided by eqn (8.2). A second is obtained by applying the mass conservation equation between the upstream state and a point within the wave. Eqn (3.30) gives

$$n_1 u_{01} = n u_0 \tag{8.3}$$

and, evaluating $u_0 = \int_{-\infty}^{\infty} uf \, \mathrm{d}\mathbf{c}$ for the distribution of eqn (8.1),

$$n_1 u_{01} = N_1 u_{01} + N_2 u_{02}. \tag{8.4}$$

The third relation is supplied by an additional moment equation. The choice of the quantity Q in this equation is arbitrary; Mott-Smith used both u^2 and u^3. With $Q = u^2$, eqn (3.27) for a steady one-dimensional flow is

$$\frac{\mathrm{d}}{\mathrm{d}x}(n\overline{u^3}) = \Delta[u^2]$$

and, using eqn (3.38) for the collision integral for Maxwell molecules,

$$\frac{\mathrm{d}}{\mathrm{d}x}(n\overline{u^3}) = \frac{3\pi}{2} A_2(5)\left(\frac{2\kappa}{m}\right)^{\frac{1}{2}} \frac{n}{m} \tau_{xx}. \tag{8.5}$$

An effective upstream mean free path may be introduced through eqn (4.55) which, for Maxwell molecules and noting that $\Gamma(7/2) = 15\pi^{\frac{1}{2}}/8$, becomes

$$\lambda_1 = \frac{16}{15\pi n_1 A_2(5)}\left(\frac{mRT_1}{\pi\kappa}\right)^{\frac{1}{2}}.$$

This may be substituted into eqn (8.5), to give

$$\frac{\mathrm{d}}{\mathrm{d}x}(n\overline{u^3}) = \frac{8}{5\pi^{\frac{1}{2}}} \frac{(2RT_1)^{\frac{1}{2}}}{n_1\lambda_1} \frac{n}{m} \tau_{xx}. \tag{8.6}$$

Now,

$$n\overline{u^3} = N_1 \int_{-\infty}^{\infty} u^3 f_1 \, \mathrm{d}\mathbf{c} + N_2 \int_{-\infty}^{\infty} u^3 f_2 \, \mathrm{d}\mathbf{c}$$

and

$$\int_{-\infty}^{\infty} u^3 f_1 \, \mathrm{d}\mathbf{c} = \frac{\beta_1^3}{\pi^{\frac{3}{2}}} \int_{-\infty}^{\infty}\int_{-\infty}^{\infty}\int_{-\infty}^{\infty} \{u_{01} + (u - u_{01})\}^3 \exp\left[-\beta_1^2\{(u-u_{01})^2 + \right.$$
$$\left. + v^2 + w^2\}\right] \mathrm{d}u \, \mathrm{d}v \, \mathrm{d}w$$
$$= u_{01}(u_{01}^2 + 3RT_1).$$

A similar expression may be obtained for the second term and, using eqn (8.3),

$$n\overline{u^3} = N_1 u_{01}(u_{01}^2 + 3RT_1 - u_{02}^2 - 3RT_2) + n_1 u_{01}(u_{02}^2 + 3RT_2).$$

Conservation of energy between the upstream and downstream states requires

$$u_{01}^2+5RT_1 = u_{02}^2+5RT_2,$$

so that,

$$\frac{\mathrm{d}}{\mathrm{d}x}(n\overline{u^3}) = \tfrac{2}{5}u_{01}(u_{01}^2-u_{02}^2)\frac{\mathrm{d}N_1}{\mathrm{d}x}. \tag{8.7}$$

Also, from eqn (1.20) and noting that τ_{xx} is symmetrical about the x axis,

$$(n/m)\tau_{xx} = n^2(\tfrac{1}{3}\overline{c'^2}-\overline{u'^2}) = \tfrac{2}{3}n^2(\overline{v'^2}-\overline{u'^2}).$$

Then, from exercise (1.5),

$$\overline{u'^2} = \overline{u^2}-u_0^2$$

and $\overline{u^2}$ may be evaluated in a similar manner to the above treatment of $\overline{u^3}$. Therefore,

$$(n/m)\tau_{xx} = \tfrac{2}{3}n(nu_0^2-N_1u_{01}^2-N_2u_{02}^2)$$

and, using eqns (8.2), (8.3), and (8.4),

$$\begin{aligned}(n/m)\tau_{xx} &= \tfrac{2}{3}\{(N_1u_{01}+N_2u_{01})^2-N_1(N_1+N_2)u_{01}^2-N_2(N_1+N_2)u_{02}^2\}\\ &= -\tfrac{2}{3}N_1N_2(u_{01}-u_{02})^2.\end{aligned}$$

N_2 may again be eliminated through eqn (8.4) to give

$$(n/m)\tau_{xx} = -\tfrac{2}{3}N_1(n_1-N_1)\frac{u_{01}}{u_{02}}(u_{01}-u_{02})^2. \tag{8.8}$$

The substitution of eqns (8.7) and (8.8) into (8.6) yields the following differential equation for N_1/n_1 as a function of x/λ_1.

$$\frac{\mathrm{d}}{\mathrm{d}(x/\lambda_1)}\left(\frac{N_1}{n_1}\right) = -\alpha\frac{N_1}{n_1}\left(1-\frac{N_1}{n_1}\right), \tag{8.9}$$

where

$$\alpha = \frac{8}{3\pi^{\frac{1}{2}}}\frac{(2RT_1)^{\frac{1}{2}}}{u_{02}}\frac{u_{01}-u_{02}}{u_{01}+u_{02}}. \tag{8.10}$$

The solution of eqn (8.9) is

$$N_1 = n_1[1+\exp\{\alpha(x/\lambda_1)\}]^{-1} \tag{8.11}$$

and the substitution of this and eqn (8.4) into eqn (8.1) provides an expression for the distribution function within the wave as a function of x/λ_1. The density profile follows directly from eqn (8.11) as

$$\frac{n-n_2}{n_1-n_2} = \frac{N_1}{n_1} = \frac{1}{1+\exp\{\alpha(x/\lambda_1)\}}. \tag{8.12}$$

The profile of any other macroscopic flow quantity across the wave may be obtained as the appropriate moment of the bimodal distribution function. The method has been extended to other inverse power law molecular models by Muckenfuss (1962).

The Mott-Smith solution has been found to be inadequate for the weaker shock waves that are adequately described by the Navier–Stokes equations, but remarkably successful in predicting the thicknesses of very strong shock waves. The experimental profiles of Schmidt do, however, show a small degree of asymmetry which is in conflict with the symmetric Mott-Smith profile. It must also be remembered that the excellent agreement for the shock thickness is based on the choice of u^2 as the quantity Q in the fourth moment equation. This is the usual choice, but there is no rigorous justification for it and Rode and Tanenbaum (1967) and Sather (1973) have shown that the result is highly sensitive to it. It may be noted that, concurrently with the experimental justification of the result with $Q = u^2$, an argument was presented (Desphande and Narasimha 1969) to the effect that the u^3 result should be superior to the u^2 result.

The BGK model equation was applied to the shock structure problem by Liepmann, Narasimha, and Chahine (1962). For one-dimensional steady flow, eqn (6.3) becomes

$$u\frac{\mathrm{d}}{\mathrm{d}x}(nf) = n\nu(f_0-f). \tag{8.13}$$

As noted in § 6.3, this is an integro–differential equation. It may be integrated to give the following integral equation

$$nf = \int_{\mp\infty}^{x} \frac{n\nu f_0(x')}{u} \exp\left(-\int_{x'}^{x}\frac{\nu\,\mathrm{d}x''}{u}\right)\mathrm{d}x', \tag{8.14}$$

where x' and x'' are dummy variables and the negative and positive signs on the lower limit apply for positive and negative u, respectively. Eqn (8.14) was solved numerically by Liepmann, Narasimha, and Chahine using an iterative scheme that required an initial guess for f_0. The results were in good agreement with the Navier–Stokes solution, and therefore with experiment, for values of $(Ma)_s$ significantly less than two. For very strong shock waves, the method led to highly asymmetric velocity and density profiles in which the upstream region was far more extended than in the experimental profiles. One possible reason for the failure of the BGK model in the upstream region of a very strong shock wave lies in the use of a collision frequency ν that is independent of f. This collision frequency was related to T through the BGK result for the coefficient of viscosity but, since this relationship is based on near-equilibrium theory, it is not adequate in the highly non-equilibrium region near the leading edge of the wave.

The finite difference solution of the Boltzmann equation using a Monte Carlo evaluation of the collision term has been applied to the shock structure problem by Hicks, Yen, and Reilly (1972). Solutions were obtained for the hard sphere molecular model for shock Mach numbers ranging from 1·1 to 10. The shock wave thickness was found to be in agreement with the Navier–Stokes theory for shock Mach numbers below two and in agreement with the Mott-Smith solution with $Q = u^2$ for shock Mach numbers above two. This latter result was in agreement with Schmidt's experimental results and with the results that had already been obtained from the direct simulation Monte Carlo method (Bird 1970*a*).

A computer program for the application of the direct simulation method to the shock structure problem may be obtained by making minor modifications to the Rayleigh problem program that is discussed in § 7.3 and listed in Appendix G. In that program, a stationary uniform gas is initially confined between a diffusely reflecting wall at $y = 0$ and a specularly reflecting wall at $y =$ YM. Then, at time $t = 0$, the wall acquires a velocity in the x direction and its temperature jumps. If, instead, the wall at $y = 0$ is specularly reflecting and acquires a velocity U_W into the gas in the y direction at zero time, a shock wave will be generated in the gas. The Mach number of this shock will be related to the velocity U_W by the standard continuum result from the Rankine–Hugoniot theory

$$\frac{U_W}{a_\infty} = \frac{2}{\gamma+1}\,\frac{(Ma)_s^2 - 1}{(Ma)_s}. \tag{8.15}$$

Here, a_∞ is the speed of sound in the undisturbed gas and, for the normalization used in § 7.3, it is equal to $(\gamma/2)^{\frac{1}{2}}$. The shock moves in the y direction and, since we chose x as the direction of motion of the shock wave in the earlier discussion of shock waves, we will now consider the program to be further modified by the interchange of the x and y axes. Because of the convergence of the walls, the NC cells into which the flow is divided will gradually be compressed, and minor modifications must be made to the coding to allow for this effect. The specular collision with the moving piston is such that the velocity of the molecule relative to the piston is reversed.

A typical result from the modified program is shown in Fig. 8.1. This shows the density contours in the distance–time plane during the formation of the shock wave. The initial density at the face of the moving wall is given by eqn (5.26) and the contours are initially tangential to the straight collisionless contours given by eqn (5.25). The effect of collisions leads to the formation

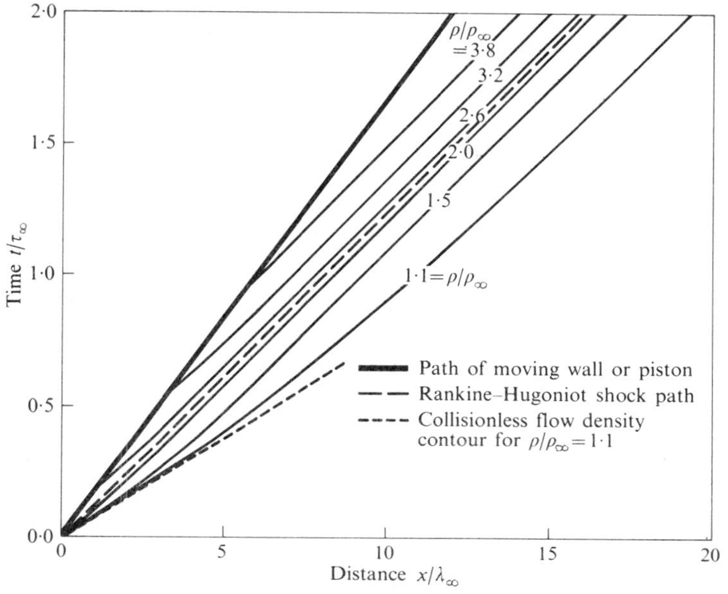

FIG. 8.1. Density contours during the formation of an $(Ma)_s = 10$ shock wave in a hard sphere gas.

of additional contours corresponding to higher densities and causes the slopes of all the contours to increase until they correspond to the Rankine–Hugoniot shock speed $U_s = (Ma)_s a_\infty$. The calculation has been terminated before the shock wave reaches the fixed wall, but it may be continued to study the shock reflection process (Bird 1969; and Diewert 1973).

The steady shock wave density profile may be deduced from the spacing of the density contours after the shock is fully formed. However, since the slope of these is known in advance from the Rankine–Hugoniot theory, the steady profile may be sampled during the running of the program by a set of sampling cells that move with the steady shock velocity. The modified program can then be used to obtain all details, ranging from the profiles of the macroscopic flow quantities to the distribution function, of steady normal shock waves in a hard sphere gas. However, in order to make meaningful comparisons with experimental measurements, it is desirable to have solutions for more general molecular models. The computational procedures for the extension of the method to include the inverse power law molecular model will now be described.

The direct simulation Monte Carlo method was outlined in § 7.1 in the framework of inverse power law molecules. Eqn (7.1) shows that the probability of a particular pair of molecules suffering a collision is proportional to $c_r^{(\eta-5)/(\eta-1)}$. An effective value of η for real gases may be inferred from a comparison of the temperature dependence of the coefficient of viscosity that is given by eqn (4.54) with the measured temperature dependence of this quantity. The most generally useful single value of η is 9 which, from eqn (4.54), corresponds to a coefficient of viscosity proportional to temperature to the power 0·75. For $\eta = 9$, the probability of collision is proportional to $c_r^{\frac{1}{2}}$ and the selection of the collision pair is similar to that outlined in the demonstration program for hard sphere molecules, except that c_r is replaced by $c_r^{\frac{1}{2}}$ in the acceptance–rejection routine. Eqn (7.2) for the time interval associated with the collision requires a value for $(\kappa/m_r)^{2/(\eta-1)}$ which, with c_r and $W_{0,m}$, fixes the effective total collision cross section defined by eqn (2.27). This is supplied by eqn (4.55) which may be used to relate κ to the mean free path in an equilibrium reference state. In the present problem, the reference state is the undisturbed upstream gas, and noting that $m_r = m/2$ in a simple gas, eqn (4.55) gives

$$\left(\frac{2\kappa}{m}\right)^{2/(\eta-1)} = \frac{2^{\frac{1}{2}}(4RT_1)^{2/(\eta-1)}}{\pi n_1 \lambda_1 A_2(\eta)\Gamma\{4-2/(\eta-1)\}}. \tag{8.16}$$

A convenient cut-off value of the dimensionless miss-distance impact parameter $W_{0,m}$ is 1·5. Eqn (2.25) shows that a fixed limit on $W_{0,m}$ applies a fixed deflection angle cut-off. This deflection angle cut-off is preferable to a fixed distance cut-off. For $\eta = 9$ and $W_{0,m} = 1{\cdot}5$, eqn (8.16) may be evaluated to give

$$\pi W_{0,m}^2(\kappa/m_r)^{2/(\eta-1)} = 3{\cdot}06(RT_1)^{\frac{1}{4}}/(n_1\lambda_1), \tag{8.17}$$

and this enables Δt_c to be evaluated for each collision from eqn (7.2). The value of W_0 for the collision ranges from 0 to 1.5 and the probability of a particular value is proportional to W_0. This probability distribution is dealt with in Appendix D and eqn (D6) gives

$$W_0 = 1{\cdot}5(R_f)^{\frac{1}{2}}. \tag{8.18}$$

The deflection angle χ in the centre of mass frame is given by eqn (2.25). This result is in the form of a definite integral with the upper limit specified by the root of an algebraic equation of order $\eta-1$. It is not practicable to solve this equation and evaluate the integral for each of the many thousands or millions of simulated collisions in a typical Monte Carlo calculation. A rational approximation to χ (in radians) for $\eta = 9$ and $W_0 \leqslant 1{\cdot}5$ is

$$\begin{aligned}\chi = \pi - 2W_0(1{\cdot}26233 + W_0(1{\cdot}84145 + W_0(-8{\cdot}87881 + \\ + W_0(20{\cdot}3313 + W_0(-23{\cdot}8155 + W_0(14{\cdot}5046 + W_0(-4{\cdot}42027 + \\ + W_0(0{\cdot}535193))))))))).\end{aligned} \tag{8.19}$$

The cut-off at $W_{0,m} = 1{\cdot}5$ corresponds to a deflection angle of approximately

1°. The azimuth angle ε is uniformly distributed between 0 and 2π, so that

$$\varepsilon = 2\pi R_{\rm f}. \tag{8.20}$$

The post-collision values of the components of the relative velocity vector then follow directly from eqn (2.22). Finally, the post collision components of the molecular velocities follow from eqns (2.1) and (2.5) exactly as in the demonstration program in § 7.3 for hard sphere molecules.

While the Maxwell molecular model is not a good representation of a real gas, it can be regarded as the limiting case of a soft molecule. Maxwell molecule computations are often required for comparison with analytical results. The collision mechanics differs from the inverse ninth power model in several respects. Firstly, all pairs of molecules are equally likely to participate in a collision and no further selection in terms of powers of $c_{\rm r}$ is required for the pair selected at random. Secondly, the term required for the evaluation of eqn (7.2) is

$$\pi W_{0,m}^2(\kappa/m_{\rm r})^{2/(\eta-1)} = 4{\cdot}39(RT_1)^{\frac{1}{2}}/(n_1\lambda_1). \tag{8.21}$$

A cut-off value of $W_{0,m} = 1{\cdot}5$ may again be used, although it corresponds to a deflection angle of 11° for Maxwell molecules and is barely adequate. Finally, the angle χ may be evaluated directly from eqn (2.32).

Bird (1970*a*) has presented results from the application of the direct simulation method to the structure of strong shock waves is a gas of inverse power law molecules, The maximum slope thickness was found to be in excellent agreement with the Mott-Smith theory for $Q = u^2$. A comparison

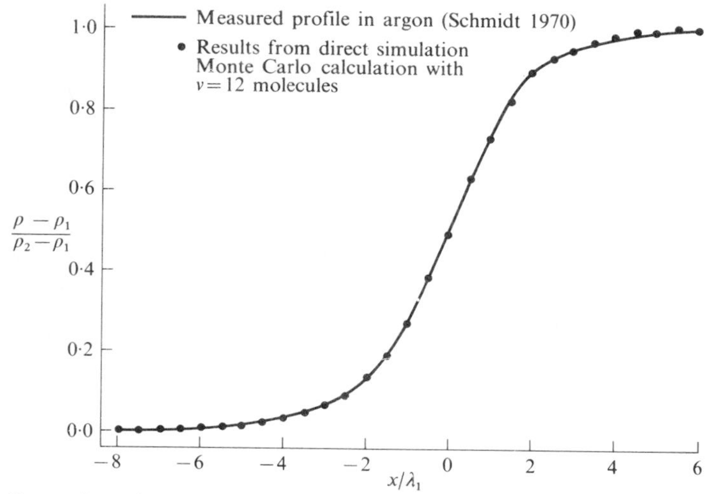

FIG. 8.2. Comparison of experimental and numerical density profiles for a normal shock wave of Mach number $(Ma)_{\rm s} = 8$.

of the calculated $(Ma)_s = 8$ profile for $\nu = 12$ molecules with Schmidt's measured profile in argon is shown in Fig. 8.2. Similar comparisons were made by Sturtevant and Steinhilper (1974) with measured profiles in other noble gases. It was found that slightly different values of η were required for the best agreement in the different gases. This was due to the basic shortcoming of the inverse power law model in that it neglects the attractive component of the intermolecular force field. Sturtevant and Steinhilper found excellent agreement between the measurements and computations when the Buckingham exponential model was used in the latter. Profiles of some of the higher moments of the distribution function are also given in Bird (1970*a*) together with a computer display representation of the distribution function. A comparison of theoretical and experimental results for the distribution function has been made by Holtz (1974). A general review of the shock structure problem has been presented by Fiszdon, Herczynski, and Walenta (1974).

8.2. Heat transfer

Consider the steady one-dimensional heat transfer in the y direction between infinite plates in the planes $y = 0$ and $y = h$. The temperatures of the lower and upper plates are T_L and T_U, respectively. This is probably the simplest problem that is available for testing the various approaches for the solution of problems in the transition regime. A large number of solutions are available and have been summarized by Springer (1971). They may be divided into those that are restricted to the small disturbance case, for which the temperature ratio T_U/T_L is very close to unity, and those that provide solutions for arbitrarily large temperature ratios. We will concentrate on the latter and commence with a description of the 'four moment solution' of Liu and Lees (1961).

As in the Mott-Smith solution for the shock wave, the assumed distribution function may be written

$$nf = N_1 f_1 + N_2 f_2 \tag{8.22}$$

but, in this case, the f_1 and f_2 describe the molecules moving in the positive and negative y directions, respectively. They are Maxwellian distributions so that, from eqn (4.1),

$$f_{1,2} = (\beta_{1,2}^3/\pi^{\frac{3}{2}}) \exp(-\beta_{1,2}^2 c'^2).$$

A further difference from the earlier distribution is that the temperatures $T_{1,2} = (2R\beta_{1,2}^2)^{-1}$, as well as $N_{1,2}$ are functions of y. For these reasons, the approximation is referred to as 'two-stream Maxwellian', rather than bimodal.

The average of any molecular quantity Q is given by eqns (8.22) and (3.3) as

$$\bar{Q} = \frac{1}{n}\left\{\frac{N_1\beta_1^3}{\pi^{\frac{3}{2}}}\int_{-\infty}^{\infty}\int_{0}^{\infty}\int_{-\infty}^{\infty} Q\exp(-\beta_1^2c'^2)\,\mathrm{d}u\,\mathrm{d}v\,\mathrm{d}w + \frac{N_2\beta_2^3}{\pi^{\frac{3}{2}}}\times\right.$$
$$\left.\times\int_{-\infty}^{\infty}\int_{-\infty}^{0}\int_{-\infty}^{\infty} Q\exp(-\beta_2^2c'^2)\,\mathrm{d}u\,\mathrm{d}v\,\mathrm{d}w\right\} \quad (8.23)$$

The boundary conditions for this problem are such that the stream velocity is zero and $c' = c$. Setting Q equal to m, v, c^2, and $\frac{1}{2}mvc^2$ yields the following expressions for the important macroscopic quantities:

$$\rho = nm = \tfrac{1}{2}(N_1+N_2)m, \quad (8.24)$$

$$v_0 = \bar{v} = (N_1/\beta_1 - N_2/\beta_2)/(2n\pi^{\frac{1}{2}}), \quad (8.25)$$

$$T = \overline{c^2}/(3R) = (N_1/\beta_1^2 + N_2/\beta_2^2)/(4nR), \quad (8.26)$$

and

$$q_y = \tfrac{1}{2}nm\overline{vc^2} = m(N_1/\beta_1^3 - N_2/\beta_2^3)/(2\pi^{\frac{1}{2}}). \quad (8.27)$$

The solution of the problem requires the determination of the four quantities N_1, N_2, β_1, and β_2. The moment equation for this problem is, from eqn (3.27)

$$\frac{\mathrm{d}}{\mathrm{d}y}\left(n\int_{-\infty}^{\infty} Qvf\,\mathrm{d}\mathbf{c}\right) = \Delta[Q]. \quad (8.28)$$

The conserved quantities m, mv, and $\frac{1}{2}mc^2$ provide obvious choices for Q in three of the four moment equations that are required. The collision integral vanishes for these equations and eqn (8.28) becomes

$$n\int_{-\infty}^{\infty} Qvf\,\mathrm{d}\mathbf{c} = \text{const.} \quad (8.29)$$

Setting $Q = m$, we obtain

$$N_1/\beta_1 - N_2/\beta_2 = 0. \quad (8.30)$$

The constant is zero in eqn (8.30) because the left hand side may be directly related through eqn (8.25) to the stream velocity normal to the plates, and the boundary conditions require this to vanish at the plates. For $Q = v$, eqn (8.29) yields

$$N_1/\beta_1^2 + N_2/\beta_2^2 = \text{const}$$

and, using eqn (8.26),

$$N_1/\beta_1^2 + N_2/\beta_2^2 = 4\,p/m \quad (8.31)$$

with the pressure $p = nmRT$ a constant for this flow. Similarly, with $Q = \frac{1}{2}mc^2$, and using eqn (8.27)

$$N_1/\beta_1^3 - N_2/\beta_2^3 = 2\pi^{\frac{1}{2}}q_y/m. \tag{8.32}$$

Since the problem essentially involves heat transfer in the y direction, Lees and Liu chose $Q = \frac{1}{2}mvc^2$ for the fourth moment. Eqn (8.28) then becomes

$$\frac{5m}{16}\frac{\mathrm{d}}{\mathrm{d}y}\left(\frac{N_1}{\beta_1^4}+\frac{N_2}{\beta_2^4}\right) = \Delta[\tfrac{1}{2}mvc^2].$$

The collision integral for $\frac{1}{2}mvc^2$ may be evaluated in a similar manner to that for u^2 in § 3.3 and, for the special case of Maxwell molecules, the result is (Vincenti and Kruger 1965, p. 364) $-\pi A_2(5)(2\kappa/m)nq_y$. The constants $A_2(5)$ and κ may be eliminated by eqns (4.54) and (4.56) for the Chapman–Enskog coefficient of heat conduction in a Maxwellian gas. The fourth moment equation then becomes

$$\frac{\mathrm{d}}{\mathrm{d}y}\left(\frac{N_1}{\beta_1^4}+\frac{N_2}{\beta_2^4}\right) = -\frac{8Rpq_y}{mK}. \tag{8.33}$$

Now, from eqns (8.26), (8.24), and (8.30),

$$T = p/(nmR) = (2R\beta_1\beta_2)^{-1} \tag{8.34}$$

which, with eqns (8.27), (8.30), and (8.31), gives

$$\frac{1}{\beta_1}-\frac{1}{\beta_2} = \frac{\pi^{\frac{1}{2}}q_y}{2p} \tag{8.35}$$

and

$$\frac{N_1}{\beta_1^4}+\frac{N_2}{\beta_2^4} = \frac{4p}{m}\left\{\frac{\pi q_y^2}{4p^2}+2RT\right\}.$$

Eqn (8.33) can therefore be written

$$q_y = -K\frac{\mathrm{d}T}{\mathrm{d}y} \tag{8.36}$$

which is, of course, the standard continuum result. The four moment solution departs from the continuum solution through the boundary conditions, which are

$$T_1 = T_{\mathrm{L}} \quad \text{at } y = 0$$

and

$$T_2 = T_{\mathrm{U}} \quad \text{at } y = h. \tag{8.37}$$

Eqns (8.34) and (8.35) enable these to be written

$$T = T_{\mathrm{L}} - \tfrac{1}{2}(\tfrac{1}{2}\pi T_{\mathrm{L}}/R)^{\frac{1}{2}} q_y/p \quad \text{at } y = 0$$

and

$$T = T_{\mathrm{U}} + \tfrac{1}{2}(\tfrac{1}{2}\pi T_{\mathrm{U}}/R)^{\frac{1}{2}} q_y/p \quad \text{at } y = h. \tag{8.38}$$

The four-moment solution is therefore equivalent to the continuum solution with a superimposed temperature jump at the plates. Eqns (4.53) and (4.56) enable the temperature jump to be expressed as a function of the Knudsen number based on the ratio of the equilibrium mean free path at temperature T to the temperature scale length. The temperature is related to the plate temperature by

$$T = T_{\mathrm{p}} \pm (75\pi/128) T_{\mathrm{p}}^{\frac{1}{2}} T^{\frac{1}{2}} \lambda \, \mathrm{d}(\ln T)/\mathrm{d}y, \tag{8.39}$$

with the positive and negative signs applying at the lower and upper plates, respectively. The temperature jump therefore decreases with the Knudsen number and the four-moment solution merges with the continuum solution as $(Kn) \to 0$. In the opposite limit of collisionless flow, the fourth moment equation becomes indeterminate as a result of T being a constant and the moment equations for the conserved quantities yield the collisionless result that was discussed in § 5.2.

The heat transfer q_y is best expressed in terms of the average density $\bar{\rho}$ between the plates and the Knudsen number (Kn) based on the ratio of the average mean free path $\bar{\lambda}$ to the plate spacing h. This must be done numerically for the four moment solution, but the procedure may be illustrated by the corresponding analysis for the continuum case. The basic continuum solution is given by eqn (5.12) and, for Maxwell molecules, the constant B is unity. Therefore

$$q_{\mathrm{c}} = -\tfrac{1}{2} C (T_{\mathrm{U}}^2 - T_{\mathrm{L}}^2)/h, \tag{8.40}$$

with the constant $C = K/T$. A mean free path may be established through the hard sphere relationship of eqn (4.53) between μ and λ. The equation of state and eqn (4.56) enable this to be written

$$\lambda = \frac{64}{75} \frac{C}{(2\pi R)^{\frac{1}{2}} p} T^{\frac{3}{2}}.$$

The continuum temperature distribution is given by the equation immediately preceding eqn (5.12) as

$$T^2 = (T_{\mathrm{U}}^2 - T_{\mathrm{L}}^2)(y/h) + T_{\mathrm{L}}^2.$$

Therefore

$$\overline{T^{\frac{3}{2}}} = \tfrac{4}{7}(T_{\mathrm{U}}^{\frac{7}{2}} - T_{\mathrm{L}}^{\frac{7}{2}})/(T_{\mathrm{U}}^2 - T_{\mathrm{L}}^2)$$

and

$$p = R\bar{\rho}\overline{T} = \tfrac{2}{3}R\bar{\rho}(T_{U}^{\frac{3}{2}} - T_{L}^{\frac{3}{2}})/(T_{U}^{\frac{1}{2}} - T_{L}^{\frac{1}{2}}),$$

so that

$$\bar{\lambda} = \frac{128}{175}\frac{C}{(2\pi)^{\frac{1}{2}}R^{\frac{3}{2}}\bar{\rho}}\frac{T_{U}^{\frac{7}{2}} - T_{L}^{\frac{7}{2}}}{T_{U}^{\frac{3}{2}} - T_{L}^{\frac{3}{2}}}. \tag{8.41}$$

The constant C may now be eliminated in eqns (8.40) and (8.41) to give

$$q_{c} = -\frac{175}{128}(\tfrac{1}{2}\pi)^{\frac{1}{2}}R^{\frac{3}{2}}\bar{\rho}(Kn)\frac{(T_{U}^{\frac{3}{2}} - T_{L}^{\frac{3}{2}})(T_{U}^{\frac{1}{2}} - T_{L}^{\frac{1}{2}})}{T_{U}^{\frac{7}{2}} - T_{L}^{\frac{7}{2}}}. \tag{8.42}$$

This continuum result may be divided by the collisionless result of eqn (5.11) to give, for Maxwell molecules,

$$\frac{q_{c}}{q_{f}} = \frac{175\pi}{512}(Kn)\frac{\{(T_{U}/T_{L})^{3} - 1\}\{(T_{U}/T_{L})^{2} - 1\}}{(T_{U}/T_{L})^{\frac{1}{2}}\{(T_{U}/T_{L})^{\frac{1}{2}} - 1\}\{(T_{U}/T_{L})^{\frac{7}{2}} - 1\}}. \tag{8.43}$$

The hard sphere case is much simpler because $\rho\lambda$ is then constant, and the ratio of the continuum to the collisionless result is

$$\frac{q_{c}}{q_{f}} = \frac{25\pi}{64}(Kn)\frac{\{(T_{U}/T_{L})^{\frac{3}{2}} - 1\}}{(T_{U}/T_{L})^{\frac{1}{2}}\{(T_{U}/T_{L})^{\frac{1}{2}} - 1\}}. \tag{8.44}$$

Numerical results for q_{y}/q_{f} have been obtained from the four moment method for the special case of $T_{U}/T_{L} = 4$ and these are plotted in Fig. 8.3

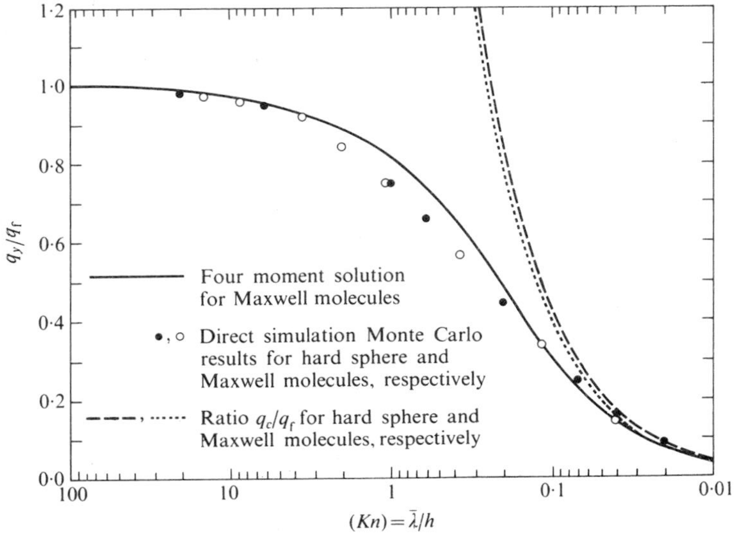

FIG. 8.3. One-dimensional heat flux for $T_{U}/T_{L} = 4$.

as a function of (Kn). As expected, these provide a smooth transition between the collisionless ($q_y/q_f = 1$) solution and the continuum solution of eqn (8.43). Results have also been obtained from the direct simulation Monte Carlo method for both hard sphere and Maxwell molecules. The demonstration program of § 7.3 is easily modified to deal with the steady heat flow problem. The major modifications are to make the upper wall diffusely reflecting at the temperature T_U and to add the necessary time averaged sampling after the establishment of the steady state. The simulation results are similar for the two molecular models and are in excellent agreement with the four moment solution as far as the bounds of validity of the collisionless and continuum regimes are concerned. There is, however, a significant difference between the two solutions at Knudsen numbers of order unity.

Fig. 8.4 shows the direct simulation result for the temperature profile between the plates for $T_U/T_L = 10$, hard sphere molecules, and a Knudsen number of 0·05. The overall heat flux is 0·90 times the continuum result. However, this flux may also be compared with the flux predicted by the application of the continuum equation (8.36) to the local temperature gradients given by the simulation. It is found that these agree to within approximately two percent, which is the order of accuracy of the simulation The difference between the overall rates is therefore due entirely to the effect of temperature jumps at the plates, as predicted by the four moment solution. The value of $(\lambda/T)(\mathrm{d}T/\mathrm{d}y)$ varies from approximately 0·13 at the cold plate to 0·075 at the hot plate, and this is just within the range for which the Chapman–Enskog (i.e. continuum) result is expected to be valid.

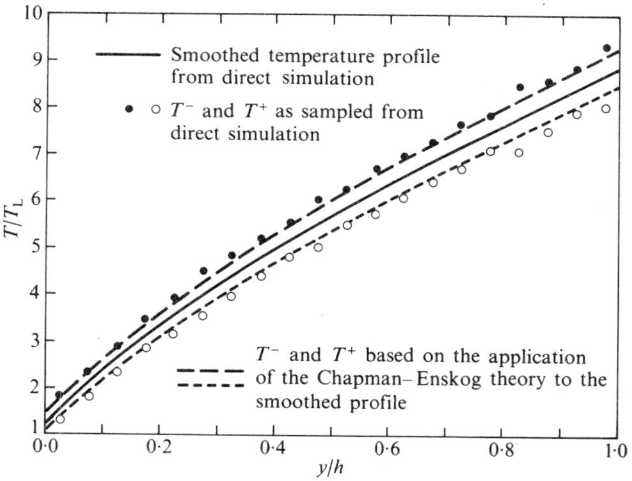

FIG. 8.4. Temperature profile for $T_U/T_L = 10$ and $(Kn) = 0{\cdot}05$.

Fig. 8.4 also provides a comparison between the temperatures T^+ and T^- of the upward and downward moving molecules, respectively, as sampled in the simulation with those based on the Chapman–Enskog distribution function of eqn (4.60) (see Exercise 4.10).

The magnitude of the temperature jump in the direct simulation result may be compared with the explicit four moment result of eqn (8.39). Although this was derived for Maxwell molecules, there is very little difference between the simulation results for Maxwell and hard sphere molecules and we will apply eqn (8.39) to the direct simulation result shown in Fig. 8.4. Using the simulation values of $(\lambda/T)\,\mathrm{d}T/\mathrm{d}y$ that were quoted above, eqn (8.39) gives $T-T_{\mathrm{L}} = 0{\cdot}267$ and $T_{\mathrm{U}}-T = 1{\cdot}28$. Fig. 8.4 shows that these are very close to the temperature jumps that were obtained in the direct simulation. For this temperature ratio, the direct simulation results and the four moment prediction of the temperature jump remain in agreement at the lower plate to a Knudsen number of approximately one. However, at the upper or hot plate, there is significant disagreement at Knudsen numbers of 0·1 and above. This is because the direct simulation temperature gradient near the upper plate becomes significantly higher than that predicted by the four moment solution. In fact, points of inflection develop in the simulation profiles at high Knudsen numbers. Significant qualitative differences therefore exist between the two results at Knudsen numbers above one, even though Fig. 8.3 shows reasonable agreement for the overall heat flux as a function of Knudsen number.

8.3. Breakdown of translational equilibrium in expansions

The breakdown of translation equilibrium in steady expansion flows and in unsteady rarefaction waves is one of the basic problems in the transition regime of gas dynamics. The effects are rather more subtle than in the shock structure problem where the both upstream and downstream conditions are fixed by the conservation equations, or in the heat transfer problem where one side of the distribution function is specified at each plate. It therefore serves as a valuable test of analytical methods, especially as theoretical results may be compared with measurements from a number of experimental studies.

The most significant analytical studies have been those of Hamel and Willis (1966) and Edwards and Cheng (1966). Both of these studies dealt with steady source flow and employed a hypersonic approximation to truncate the moment equations. The analyses differed in their treatment of the collision term in that Edwards and Cheng used the BGK model, while Hamel and Willis used a Maxwell molecule evaluation of the required collision integrals. Both approaches lead to a similar set of moment equations. These predict that the breakdown of equilibrium leads to a gradual separation of the kinetic temperature T_r based on the molecular velocity components parallel to the flow from the kinetic temperature T_n based on the velocity components normal to the flow. The parallel temperature gradually approaches a constant or frozen value, as would be expected for a collisionless flow, but the normal temperature is predicted to decline inversely with the radius r. This r^{-1} temperature dependence is less than the $r^{-\frac{4}{3}}$ dependence for an isentropic spherical expansion of a monatomic gas, whereas the expected collisionless result is an r^{-2} dependence due entirely to the geometric effect. Hamel and Willis pointed out that, since $T_r \gg T_n$ at large radii, a very small amount of collisional transfer can produce large effects on T_n. Edwards and Cheng investigated the distribution function corresponding to the far field T_n and found that most of the energy is in the tails of the distribution function.

The direct simulation Monte Carlo method has been applied to the steady source flow problem by Bird (1970*b*). The throat Knudsen number $(Kn)^*$ is defined for the flow as the ratio of the mean free path in the stagnation region to the radius of the throat r^*. The results are best illustrated by a plot of temperature against radius. The temperature is normalized by the stagnation temperature T_0 and the radius by the throat radius r^*. The result for hard sphere molecules and $(Kn)^* = 0{\cdot}002$ is shown in Fig. 8.5(a). This gives qualitative support to one of the major predictions of the approximate

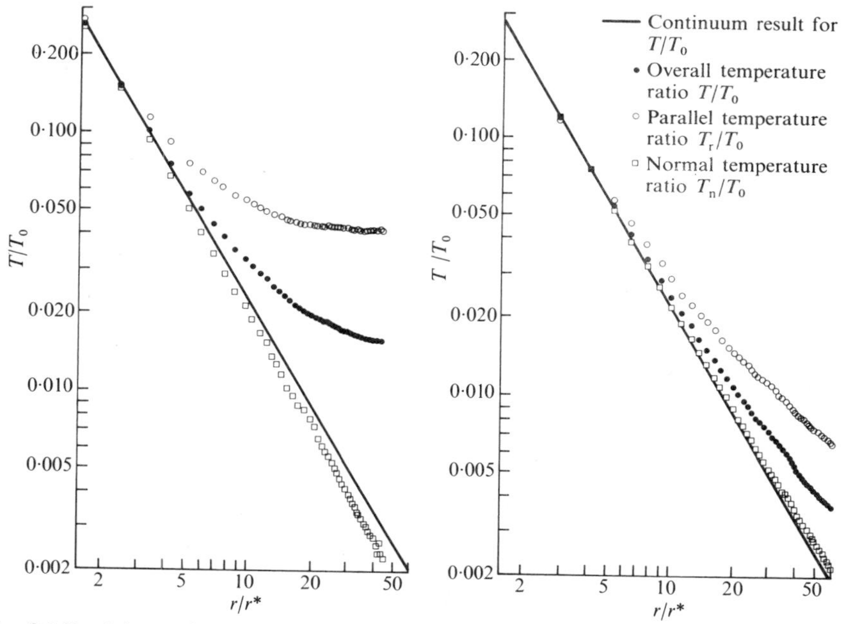

FIG. 8.5. Breakdown of translational equilibrium in a steady spherical expansion with $(Kn)^* = 0{\cdot}002$. (a) Hard sphere gas. (b) Maxwell gas.

analyses—that the freezing of the parallel kinetic temperature occurs gradually over a wide range of radius and Mach number. However, in the case of the normal temperature, there is a qualitative difference between the simulation result, which shows this temperature falling below the $r^{-\frac{4}{3}}$ line, and the moment solutions which predict an r^{-1} decay. The simulation calculation was repeated for a gas of Maxwell molecules, with the result shown in Fig. 8.5(b). The freezing of the parallel temperature is much less rapid than for the hard sphere molecules and the frozen state is not approached in this calculation. The most interesting change is that the normal temperature diverges above, rather than below, the $r^{-\frac{4}{3}}$ continuum line, but the rate of decay is still significantly greater than the r^{-1} moment prediction. However, a 9° deflection angle cut-off was applied in the computation and, although this is satisfactory in most applications, the neglect of the very large number of weak interactions might have produced a significant error in this situation. Also, assuming that the far-field transfer is largely contained in the tails of the distribution function, there may be some doubt whether the comparatively small sample in the Monte Carlo simulation is adequate. The time-averaged sample at each of the points in Fig. 8.5 is approximately 5000 and this provides a reasonable sample of high speed molecules. Also, the computational details such as cell size and sample number were identical in the two cases and, since there were sufficient in the Maxwell molecule case to keep the T_n curve above the $r^{-\frac{4}{3}}$ line even with the 9° cut-off, the falling of the hard sphere curve below this rate is almost certainly significant. The rate of decay of T_n in a real monatomic gas would be expected to be between the Maxwell and hard sphere rates and might therefore be expected to be very close to the isentropic rate. Measurements by Muntz (1967) and by Cattolica, Robben, Talbot, and Willis (1974) have in fact, led to points that lie almost exactly on the $r^{-\frac{4}{3}}$ isentropic curve.

One of the objectives of the direct simulation study was to establish a criterion for the onset of non-equilibrium and this ideally requires the determination of a parameter which correlates results for all Knudsen numbers, molecular models, and flow geometries. It was found that this requirement was satisfied by the parameter

$$P = (1/\nu)|D(\ln \rho)/Dt|, \tag{8.45}$$

which may be written for a steady one-dimensional flow in the x direction as

$$P = \{u/(\rho\nu)\}|\mathrm{d}\rho/\mathrm{d}x|. \tag{8.46}$$

The collision frequency ν may be related to the coefficient of viscosity through eqns (8.45) and (8.46). The ratio T_r/T_n was plotted against P for all the simulation cases and this indicated that the value of P for breakdown is approximately 0·04. More recently, Cattolica *et al.* (1974) have presented experimental results for the onset of non-equilibrium. All results for T_r/T_n were found to collapse to a single curve when plotted against P, and the value of P at the onset of non-equilibrium was found to be in good agreement with the value predicted by the simulation study. The general behaviour of the parameter P in steady cylindrical and spherical expansions has been studied by Bird (1970*b*). The parameter decreases to a minimum at a Mach number close to 1·5 and then increases. For very large Mach numbers, P is proportional to $(Ma)^n$, where n varies from 0 for the cylindrical expansion of Maxwell gas to 2·5 for a spherical expansion of a hard sphere gas. For breakdown occurring in a hypersonic flow at a constant value of P, the analysis shows that the breakdown Mach number would be expected to be equal to a constant multiplied by the throat Knudsen number raised to the power $-1/n$. The value of this power is in agreement with the value predicted by Hamel and Willis (1966).

Bird (1970*b*) also applied the direct simulation Monte Carlo method to the breakdown of translational equilibrium in a complete, one-dimensional, unsteady rarefaction wave in a hard sphere gas. It was found that the ratio of the parallel and normal kinetic temperatures was a function of the initial position of the fluid element, but was independent of time. An analysis of this flow showed that the parameter P remains constant along particle paths in this flow. This is entirely consistent with the numerical result that the T_x/T_n curves are independent of time. It was found that translational equilibrium breaks down for gas elements that are initially within fifteen undisturbed mean free paths of the origin of the rarefaction. The value of p on the particle path that is initially fifteen mean free paths from the centre is 0·04, and the quantitative agreement with the steady flow result gives further support to the use of P as the correlating parameter.

8.4. The Rayleigh problem

We have already met the Rayleigh problem in the context of collisionless flows in § 5.6 and as the subject of the demonstration program for the direct simulation Monte Carlo method in § 7.3. The output from a typical application of the Monte Carlo simulation is reproduced after the listing of the demonstration program in Appendix G. The data for this was chosen to allow a direct comparison with the finite difference solution of the BGK equation for this problem by Chu (1967). At zero time, the plate velocity is set equal to twice the most probable molecular speed in the undisturbed gas, while the wall temperature is set to 1·6 times the temperature of this gas.

The surface quantities have been printed at successive intervals of one mean collision time in the undisturbed gas (which is numerically equal to $\pi^{\frac{1}{2}}/2$ for the normalization that has been employed). The initial values of these

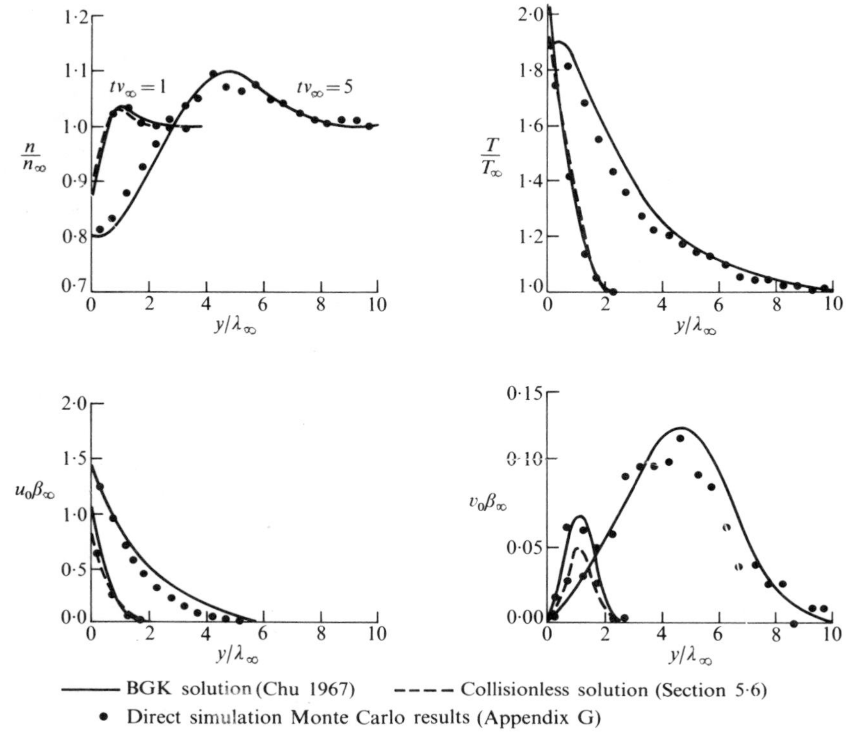

FIG. 8.6. Contours of number density n/n_∞, temperature T/T_∞, parallel velocity $u_0\beta_\infty$, and normal velocity $v_0\beta_\infty$ at several instants of time in the Rayleigh flow of a hard sphere gas with $U_w = 2/\beta_\infty$ and $T_w = 1{\cdot}6T_\infty$.

quantities should be equal to the time independent collisionless quantities. For this data, eqns (4.20) and (5.57) to (5.58) give the following results:

$$N\beta_\infty/n_\infty = 0{\cdot}2821;$$

$$p_i\beta_\infty^2/\rho_\infty = 0{\cdot}25;$$

$$p_r\beta_\infty^2/\rho_\infty = 0{\cdot}3162;$$

$$\tau_i\beta_\infty^2/\rho_\infty = 0{\cdot}5642;$$

$$\tau_r\beta_\infty^2/\rho_\infty = 0;$$

$$q_i\beta_\infty^3/\rho_\infty = 0{\cdot}8463;$$

and

$$q_r\beta_\infty^3/\rho_\infty = 0{\cdot}4514;$$

where the subscript ∞ denotes the undisturbed gas. These values are in excellent agreement with the extrapolations to zero time of the simulation results. Note that the number flux shows a maximum at between one and two mean collision times. The incident shear stress and heat flux are the quantities most affected by the collisions in the gas.

The flow field properties have been printed at intervals of one undisturbed mean collision time, up to five mean collision times. The first and final results for the number density, temperature, parallel velocity, and normal velocity are compared with the BGK results in Fig. 8.6. The collisionless results from § 5.6 are also included for one mean collision time. Since the collisionless results are functions of y/t, the collisionless results for $tv = 5$ could be obtained from that for $tv_\infty = 1$ by the simple expedient of multiplying the height y/λ_∞ by five. It is obvious that the collisions have had a strong effect on the flow at $tv_\infty = 5$. The BGK solution and the direct simulation Monte Carlo results are in good general agreement. It may be noted that the perturbations of the normal velocity is comparatively small and that, although the statistical scatter becomes significant, the direct simulation method is capable of providing useful results for perturbations of the order of five to ten per cent.

The program in Appendix G was executed on a CDC Cyber 72 computer and required just on 1000 seconds of central processor time.

9

MULTI-DIMENSIONAL FLOWS

9.1. General approaches

FEW of the transition regime approaches that lead to useful solutions for one-dimensional flows have proved capable of extension to flows with more than one spatial dimension. The major factor is that addition of even one dimension in physical space causes the distribution function to become three-dimensional, rather than axially symmetric, in velocity space. The moment method depends on an intuitive choice of an assumed distribution function and it is hardly surprising that attempts to extend it to flows with more than one spatial dimension have generally proved to be unsuccessful. The treatment of some limiting cases by Grad and Hu (1969) appears to be one of the few exceptions, but their approach is restricted to simple flow geometries. A similar difficulty arises with any method that, in common with the finite difference and test particle Monte Carlo methods, requires an initial guess of the complete flow field in order to commence an iterative procedure that hopefully converges to a unique solution. Quite apart from the difficulty in choosing a suitable initial flow configuration, the application of the finite difference Monte Carlo method to two-dimensional flows appears to be computationally impracticable. The existing applications of this method to one-dimensional flows have been made with the three-dimensional phase space divided into approximately 230 cells. A vastly increased number would be required to cope with the five phase space dimensions associated with a two-dimensional flow. It must also be remembered that velocity space is unbounded and, as the geometry becomes more complex, it is increasingly difficult to decide where to place the outer limits that must be specified for any system that deals with a finite number of cells in velocity space. As with the one-dimensional flows, we will not consider methods which are applicable only to flows involving small perturbations of some undisturbed state, or to small departures from the limits of collisionless and continuum flow. This leaves only the model equation approach and the direct simulation Monte Carlo method.

The BGK and similar model equations have had very restricted application to nonlinear two-dimensional transition regime flows. This is because the application of these models leads to a set of equations that are generally more difficult than the Navier–Stokes equations for viscous compressible continuum flow. Finite difference solutions of the Navier–Stokes equations for representative two-dimensional problems that include the merging of the shock wave structure with the boundary layer profile have only recently been obtained. Examples are the blunt body calculations of Victoria and Widhopf (1973) and the flat plate leading edge study by Tannehill, Mohling, and Rakich (1974). The latter study employed a slip boundary condition and the results exhibited an inflection in the density profile normal to the surface. This had been observed in experiments (Metcalf, Lillicrap, and Berry 1969) and predicted by Monte Carlo simulation studies (Vogenitz, Broadwell, and Bird 1970), but had thought to be inaccessible to the Navier–Stokes equations. Application of the full Navier–Stokes equations to flows of more complex geometry is a very difficult task and the application of the BGK equation would be a truly formidable one. Also, we have seen that there is evidence that the assumptions inherent in the BGK formulation break down in highly non-equilibrium regions such as the leading edge of a strong shock wave.

The direct simulation Monte Carlo method was first applied to flows about bodies of simple shape (Vogenitz, Bird, Broadwell, and Rungaldier 1968). A typical result from a more recent application by Crawford and Vogenitz (1974) is shown in Fig. 9.1. This deals with the two-dimensional supersonic flow past a cylinder. The velocity profiles along the stagnation streamline are shown for a range of Knudsen number from 0·02 to 1. The

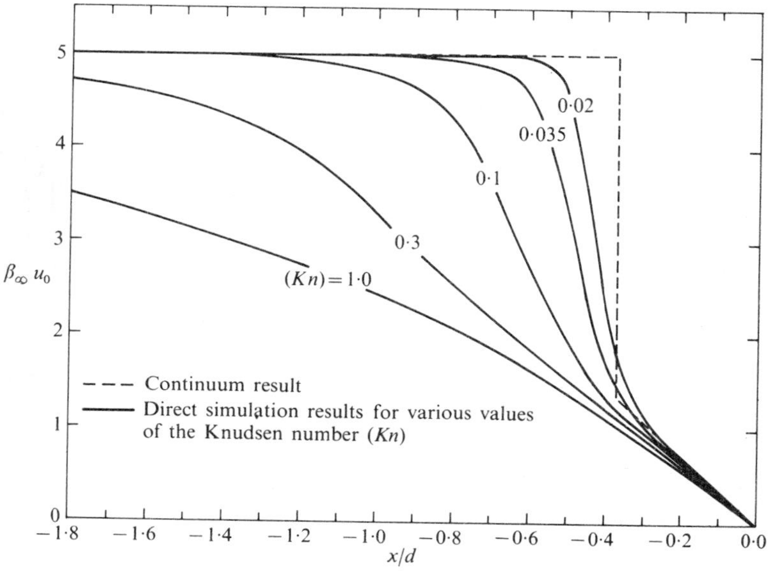

FIG. 9.1. Velocity profiles along the stagnation streamline in front of a circular cylinder in an inverse twelfth power gas at a speed ratio of five.

Knudsen number (Kn) is equal to the ratio of the undisturbed mean free path λ_∞ to the diameter d. The x axis is in the streamwise direction and the origin is at the centre of the cylinder. The calculation was made for inverse twelfth power molecules, at a speed ratio of five, and for zero heat transfer to the cylinder. The continuum result shows the Rankine–Hugoniot shock wave at the accepted stand-off distance together with a linear variation of velocity from the rear of the shock to the stagnation point.

The programs for the flows past bodies of simple shape are comparatively straightforward extensions of the one-dimensional programs that have been described in the earlier chapters. As in the one-dimensional programs, the molecules are indexed to a network of spatial cells. The geometry of these cells is generally related to that of the body; analytical expressions being required for both the body and the cell boundaries. This effectively restricts the method to flows of simple geometry and necessitates considerable re-programming whenever the geometry is changed. In order to overcome this problem, an alternative scheme has been developed in which the molecules are indexed to a network of points rather than areas. This has permitted the development of a 'universal program' that may be applied to two-dimensional or axially symmetric flows of arbitrary geometry, merely through changes in the data input. The basic routines for this approach are described in the following section, together with the additional routines that are required for any two-dimensional or axially symmetric flow.

9.2. Simulation procedures for two-dimensional and axially symmetric flows

A two-dimensional or axially symmetric flow field is divided into an arbitrary number of regions similar to those shown in Fig. 9.2. The sides of each region are numbered from one to four and can form part of a solid surface, a boundary at which the incoming gas molecules are specified, an axis or plane of symmetry, or an interface with another region. The positions of the four sides are defined by the coordinates of an equal number of points along sides one and three. These coordinates are read as data and the sides are the lines of best fit to the points. Corresponding points along sides one and three are joined by straight lines, with the outer pair defining sides two and four. Each of these lines is divided into segments; the number and the relative sizes of which are prescribed in the data. The lines joining the corresponding divisions between segments are of the same family as lines one and three. The two sets of lines divide the region into a network of quadrilateral cells.

The centroids of the quadrilaterals define the 'interior grid points'. One 'exterior grid point' is defined for each segment that lies on a side of the region as the image of the adjacent 'interior grid point' when it is reflected in the segment. The grid points rather than the array of lines define the cells,

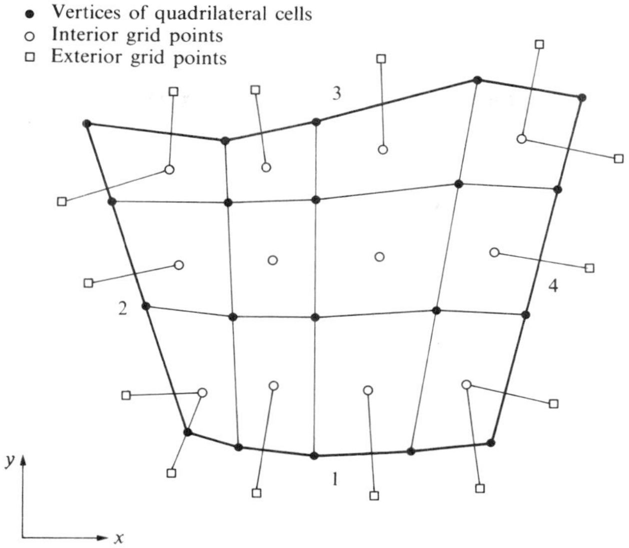

FIG. 9.2. A typical region in a two-dimensional flow field.

in that a molecule is said to be in a given 'cell' when it is closer to that grid point than to any other. Since the grid points lie in a well defined array of 'rows' and 'columns', their locations may be stored and indexed in an easily referenced manner.

When a molecule moves to a new location, it is a simple matter to determine whether it has moved closer to one of the adjacent points. This is generally a 'four point process' in which only the points along the same row or column as the current points are checked. Should the molecule have moved closer to one of the four adjacent points, this point becomes the current point and the process is repeated. This procedure is exact as long as the grid of points is rectangular. However, in a highly oblique and irregular grid, the molecule may be closest to a point that is diagonal to the current point, but is not transferred to it because it is closer to the current point than to the points along the same row and column. This means that, unlike the rectangular grid case, the effective cell boundaries cannot be precisely defined and the exact cell area cannot be calculated. For most applications, satisfactory results are obtained if the consequent fuzziness of cells and boundary elements is accepted and the cell areas are based on the original network of quadrilateral cells. Should it be necessary to have exact procedures for an oblique cell network, an 'eight point process' may be adopted. In this, the position of the molecule is checked against all eight adjacent points, including the diagonal points. It permits the exact calculation of effective cell boundaries for all configurations.

As long as a molecule is nearest one of the interior grid points, it is within the region. If it moves closest to one of the exterior grid points, it must have crossed the adjacent boundary segment and the appropriate boundary interaction may be calculated.

A given flow may consist of an arbitrary number of regions. The orientation of the regions relative to the numbering of the sides is quite arbitrary and a given side may interface with more than one adjacent region. The geometrical configuration of the flow is defined by the coordinates of the points along sides one and three of each region; no analytical descriptions are required for the boundaries. Although sides two and four must be straight it is possible to devise arrangements of regions to cope with all flow geometries.

The program follows the flow chart of Fig. 7.1 and, apart from those for the cell system, very few new routines are required. One of these is for the generation of entering molecules from the equilibrium distribution function for a moving gas. The inward number flux across an element of area is given by eqn (4.18). Also, the separation of variables that led to this equation from eqn (4.17) shows that the velocity component normal to the entry plane is the only one that requires special attention. The thermal velocity components in the other directions are generated as for a stationary equilibrium gas (see Appendix D); the overall velocity components being obtained from these by the addition of the appropriate stream velocity components. The distribution function for the thermal velocity component u'_n normal to the boundary element also follows from these equations. It is,

$$f_{\beta u'_n} = \pi^{-\frac{1}{2}}\beta^{-1}(\beta u'_n + s_n)\exp(-\beta^2 u'^2_n) \tag{9.1}$$

where s_n is the component of the stream velocity normal to the boundary element. The acceptance–rejection method must be used for the sampling from this distribution also. The maximum of the distribution occurs when $\beta u'_n = \{(s_n^2+2)^{\frac{1}{2}} - s_n\}/2$ and the normalized distribution is

$$f'_{\beta u'_n} = \frac{2(\beta u'_n + s_n)}{s_n + (s_n^2+2)^{\frac{1}{2}}}\exp\left[\frac{1}{2} + \frac{s_n}{2}\{s_n - (s_n^2+2)^{\frac{1}{2}}\} - \beta^2 u'^2_n\right]. \tag{9.2}$$

The procedure is then to choose a value of $\beta u'_n$ that is uniformly distributed between, say, -3 and $+3$, but subject to the condition that $\beta u'_n + s_n$ must be positive. The value of $f'_{\beta u'_n}$ is then evaluated and is compared with the next random fraction R_f. The value of β is that appropriate to the entering gas, and the normal velocity component of the entering molecule is $(\beta u'_n + s_n)/\beta$.

The three velocity components u, v, and w are stored for each molecule together with two position coordinates, say x and y. In the case of an axially symmetric flow with the axis in the x direction, y becomes the radius. If a molecule is at radius y with velocity components v_1 and w_1, and moves for time Δt, the new radius is

$$y = \{(y_1 + v_1 \Delta t)^2 + (w_1 \Delta t)^2\}^{\frac{1}{2}}. \tag{9.3}$$

Also, the v and w velocity components must be rotated so that v remains the radial component. The new values of v and w are

$$v = \{v_1(y_1 + v_1 \Delta t) + w_1^2 \Delta t\}/y$$

and (9.4)

$$w = \{w_1(y_1 + v_1 \Delta t) - v_1 w_1 \Delta t\}/y.$$

A two-dimensional flow may simply be regarded as having unit width, so that the cell volume is numerically equal to its area in the x, y plane. On the other hand, the volume of a cell in an axially symmetric flow must be calculated from the appropriate annulus. The volumes of cells located at large distances from the axis are therefore large in comparison with the volumes of cells of similar cross-sectional area near the axis. This means that, if the density is similar in each cell, the number of simulated molecules, and consequently the sample size for the macroscopic properties, is either excessively small near the axis or excessively large well away from the axis. A small sample leads to a large degree of statistical scatter in the results, while a large sample leads to excessive computation time. The answer to this problem lies in the use of *weighting factors*.

Each simulated molecule may be regarded as being representative of some very much larger number of molecules in the real gas. There is no reason why the ratio of real to simulated molecules should be the same in all parts of the flowfield. This is the principle behind the use of weighting factors and, since the weighting factors are to be used here to compensate for large disparities in cell volume, they are most conveniently assigned to each cell. If the weighting factor for cell n is W_n, each molecule in this cell represents W_n weighted molecules. The weighting factor is conveniently normalized such that the minimum value is unity. The maximum weighting factor is not necessarily associated with the smallest cell, because weighting factors may be used to compensate for large variations in density instead of (or in addition to) large variations in cell volume.

Whenever weighting factors are used, care must be taken in all procedures to distinguish between the actual and weighted number of molecules. It is the weighted number density that appears in the various equations, but the actual number that must be allowed for in the allocation of storage. The number N_{M} of simulated molecules in a flowfield of N_{c} cells is

$$N_{\mathrm{M}} = \sum_{n=1}^{N_{\mathrm{C}}} (n_n V_n / W_n), \tag{9.5}$$

where V_n and n_n are, respectively, the volume and number density in cell n. A frequent requirement is for the determination of the number density n_∞ in an initially uniform flow containing N_{M1} molecules. This follows from eqn (9.5) as

$$n_\infty = N_{M1}\left\{\sum_{n=1}^{N_c} (V_n/W_n)\right\}^{-1}. \tag{9.6}$$

In general, a flowfield has open boundaries and, even without weighting factors, provision must be made for a systematic change in the number of simulated molecules during unsteady processes and for fluctuations during macroscopically steady processes. When a molecule moves from a cell with weighting factor W_n to one with weighting factor W_m, provision must be made to either remove or duplicate the molecule. The number of molecules in the new cell should be equal to W_n/W_m and, since this will generally not be an integer, the removal or duplication is based on an acceptance–rejection procedure. A listing of an efficient FORTRAN routine for handling the variable number of simulated molecules is presented in Appendix H.

9.3. Three-dimensional flows

The extension of a two-dimensional method to deal with three-dimensional flows generally requires very little that is new in principle. It does, however, require a significant increase in computing time and storage capacity, particularly the latter. In the case of the direct simulation Monte Carlo method, the molecular motion and collisions are always treated as three-dimensional phenomena. The only additional requirement for a three-dimensional flow is that a record must be kept of the molecular positions in all three spatial directions, rather than in just one or two. If all the boundaries in a particular problem are sufficiently simple to be described analytically, the molecules may be indexed to a set of volumetric cells. For flows of complex geometry, it is necessary to extend the point indexing method to three dimensions.

The logical extension of the two-dimensional point indexing scheme is to divide the three-dimensional field into regions with six faces, numbered one to six as in Fig. 9.3. Each face defines either a solid surface, a plane of symmetry, a boundary with specified inflow conditions, or an interface with another region.

The coordinates of a network of points on faces one and three are read in as data. There is an equal number of points on each face and they are arranged in rows and columns. The points on face one are joined to the corresponding points on face three by straight lines. Therefore, while faces one and three may have double curvature, the other faces are defined by the straight lines joining the points at the edges of the array of rows and

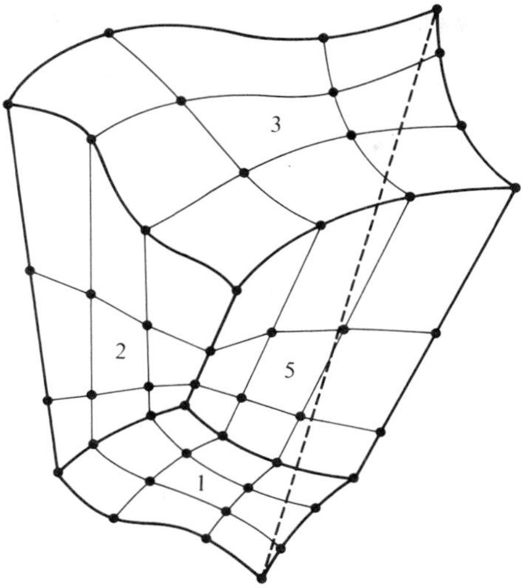

FIG. 9.3. A typical region in a three-dimensional flow field. (Faces 4 and 6 of the region are hidden.)

columns. All the straight lines are divided into a set ratio, thus defining a set of cells that will be described as 'hexahedral' even though the four points defining the vertices of each face do not necessarily lie in a plane. Since these cells are arranged in a systematic and easily indexed array of rows, columns, and layers, they may be used as reference volumes for the flow properties.

The centroids of the hexahedral cells could be used as the interior grid points for monitoring the motion of the molecules. However, since the faces of these cells are not necessarily plane, it would not be possible to define the boundaries through the use of image points as exterior grid points. The solution is to establish a finer grid through the subdivision of each hexahedral cell into five tetrahedral elements. The division of a given hexahedral cell into tetrahedral elements may be done in two ways, thus leading to type A and type B cells as shown in Fig. 9.4. In both cases, there are four 'external' tetrahedra numbered one to four and the 'internal' number five tetrahedron. The network of hexahedral cells is divided into tetrahedral elements such that type A and type B divisions alternate in all three directions; that is along rows, columns, and layers. The two triangular faces of external tetrahedral cells that form each face of a hexahedral cell then coincide with the corresponding triangular faces in the neighbouring hexahedral cells. In fact, the tetrahedral elements may be numbered in such a way that the coinciding faces always belong to terahedral elements having the same code number.

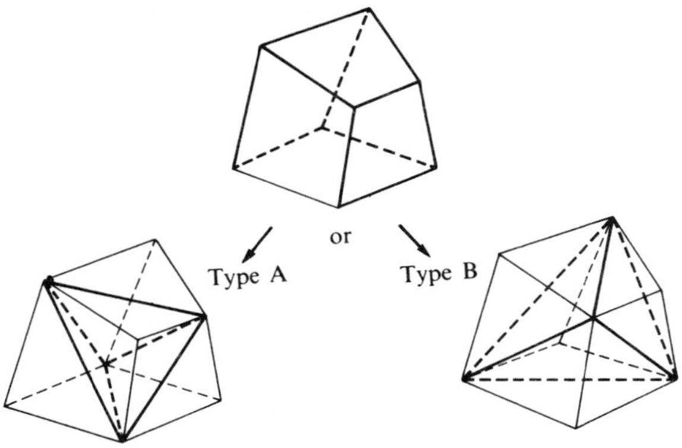

FIG. 9.4. Alternate ways in which a deformed hexahedron with six non-planar faces may be divided into five tetrahedra.

The centroids of the tetrahedral elements constitute the interior grid points. A given molecule is then indexed by the code number of the tetrahedral element, together with the code number of the hexahedral cell of which the element forms a part. A given molecule is said to be in a given cell and element when it is nearer to the centroid of that element than to any other grid point. This must be tested by reference to the four adjacent grid points. The four elements adjacent to an internal tetrahedral element are the four external elements in the same cell. Of the four elements adjacent to any particular external tetrahedral element, one is the internal element of the same cell, but the others lie in adjacent hexahedral cells. It is most important that the indexing should be computationally efficient; this is achieved if advantage is taken of the abovementioned possibility of arranging the numbering system such that all adjacent external elements have the same code number. When a triangular face of a tetrahedral element also lies on the face of a region, an exterior grid point is defined by the image of the centroid when it is reflected in the triangular face. The boundaries of the flow are therefore defined by a continuous array of plane triangular elements.

9.4. Range of application

The low density of high Knudsen number limit to the range of application of the Navier–Stokes equations is set by the breakdown in the validity of the Chapman–Enskog theory for the transport properties. The high density of low Knudsen number limit to the application of the direct simulation Monte Carlo method is set by the magnitude of the computing requirements. While the limit on the Navier–Stokes approach is fixed, that on the direct simulation approach may be expected to be progressively extended by advances in computer capabilities. There is already a considerable overlap of the ranges of application of the two methods. It is usually assumed that, when both the microscopic and macroscopic approaches are valid, there is an overwhelming case in favour of the macroscopic or continuum approach. This assumption is based on the difficulties that are traditionally associated with analytical or numerical solutions of the Boltzmann equation, as compared with similar solutions of the Navier–Stokes equations. We will now determine whether this assumption requires any modification when the direct simulation Monte Carlo approach is taken into account.

It should be made clear at the outset that consideration will be limited to flows requring numerical methods for the continuum solution. An analytical solution would obviously be preferred to a numerical solution, irrespective of whether the latter is based on the continuum or particle model. Also, if the disturbances in the flow are sufficiently small to allow a small-perturbation solution, the problem will generally be inaccessible to the direct simulation method. This is because of the difficulty in distinguishing between weak disturbances and the statistical scatter due to the finite sample size. Note that the stream velocity is the average of molecular velocities of the order of the speed of sound, so that the statistical scatter in this quantity will be relative to the speed of sound. The direct simulation method is, therefore, unlikely to provide useful solutions for low subsonic flows.

In view of the above considerations, the macroscopic and microscopic methods will be compared for a test flow that involves large disturbances and requires numerical analysis for the continuum solution. This is the two-dimensional steady flow field produced by the impingement of an under-expanded jet on a wall. This is a mixed subsonic–supersonic flow and provides a severe test for numerical methods since it contains a shear layer at the edge of the jet, a boundary layer along the surface, and an embedded shock wave at an unknown location.

A typical result from the direct simulation Monte Carlo method for the Mach number contours in the test flow is shown in Fig. 9.5. The flow is symmetric about the x axis and a sonic jet of height H is located at $x = 0$. This jet discharges in the x direction into a static atmosphere at an ambient temperature equal to the jet stagnation temperature; the ambient pressure being equal to 0·015 times the jet stagnation pressure. The flow in Fig. 9.5 is for an ambient mean free path equal to $0{\cdot}15H$ which corresponds to a mean free path at the jet exit equal to approximately $0{\cdot}0012H$. The calculation was made for a diatomic gas using the energy sink model that is described in § 11.3. The 'universal' program of § 9.2 was used for the calculation. The flow field was divided into three regions with a total of 460 internal cells. This means that a typical cell dimension was much larger than the local mean free path, but was generally small compared with the scale length of the local flow gradients. The exception to this was in the interior of the

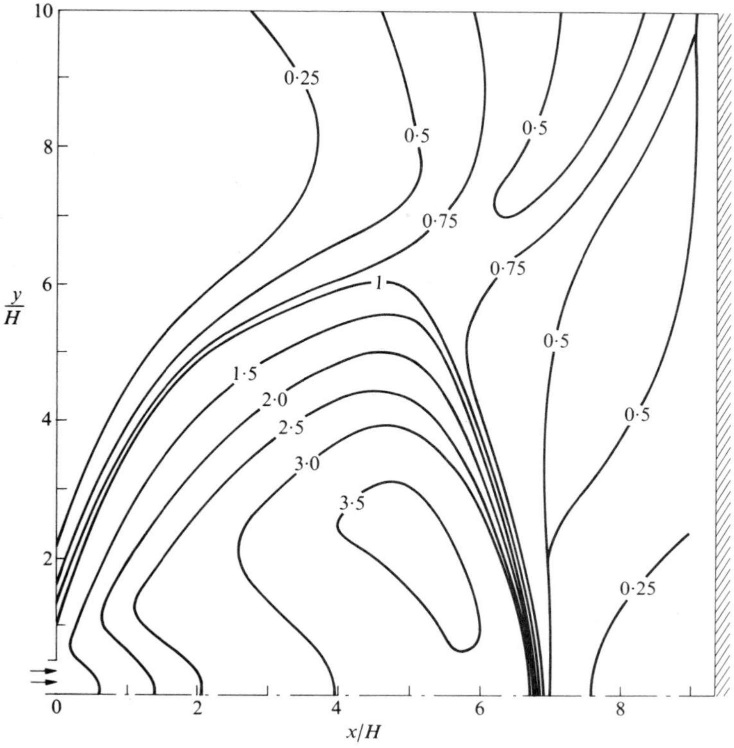

FIG. 9.5. Mach number contours in the impingement of an underexpanded jet plume on a wall.

shock wave; the width of this being fixed by the cell size rather than by the local mean free path. The simulation commenced with the flow field set up as undisturbed ambient gas and the jet was instantaneously started at zero time. The time-averaged sampling of the macroscopic quantities was commenced after the flow field had become steady.

The inviscid continuum flow corresponding to that in Fig. 9.5 has been computed by Sinha, Zakkay, and Erdos (1971) by a time dependent finite difference method. It should be noted that this is an artificial time dependence and the flow was not followed through a physically real unsteady process. The only significant difference between the continuum flow field and that of Fig. 9.5 is that the latter includes a thick boundary layer along the wall. The Mach number distributions along the centreline of the flow are compared in Fig. 9.6. The artificial viscosity in the continuum finite difference method leads to a shock wave that is spread over several mesh widths. The apparent shock thickness is therefore of the same order as that given by the simulation method, with the pre-shock and post-shock oscillations

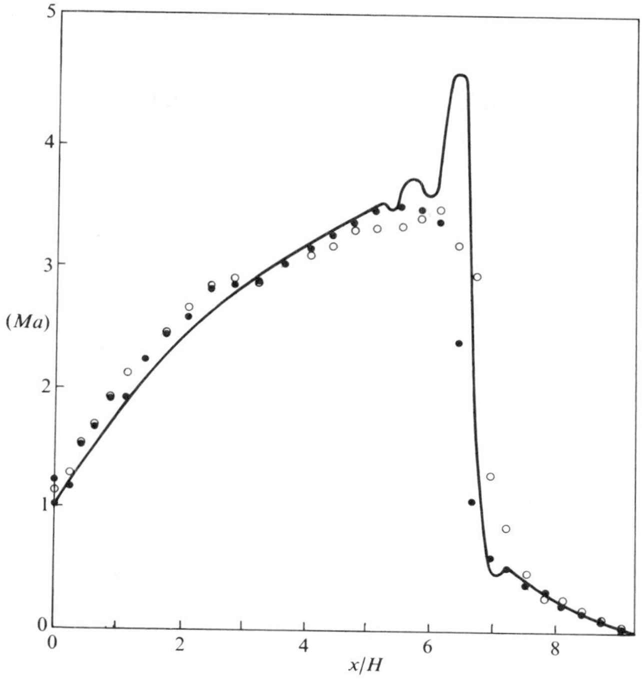

FIG. 9.6. Mach number distribution along the centreline of the jet impingement flow. —— Continuum finite difference calculation (Sinha *et al.* 1971); ● Direct simulation result for diffuse reflection and $\lambda_\infty = 0{\cdot}05H$; ○ Direct simulation result for specular reflection and $\lambda_\infty = 0{\cdot}15H$.

constituting an added disadvantage. The shock position in the simulation is slightly upstream of the continuum shock position, but this is due to the displacement effect of the boundary layer along the wall. This boundary layer is a consequence of the specification of diffuse reflection at the wall and is effectively removed if specular reflection is specified instead of diffuse. A run was made with specular reflection and this resulted in exact agreement on the location of the shock wave at the axis. The shock location and the pressure distribution along the wall were also found to be in agreement with experiment. Further details of this flow computation are given in Bird (1973).

Isentropic flows and flow regions are most easily modelled since, once sufficient collisions are computed to keep the flow in equilibrium, the flow will not be altered by the computation of additional collisions and is therefore independent of the Knudsen number. The specification of specular reflection at solid surfaces enables the simulation method to deal efficiently with flows that are inviscid in the sense that there are no boundary layers on solid surfaces, but viscous in that shock waves are properly accounted for. The comparative computation times for the jet impingement problem indicate that the simulation method does not necessarily require more computer resources than the continuum finite difference approach. The computational task becomes formidable only when it is necessary to simulate a boundary or shear layer at relatively high densities and Reynolds numbers. This task may be alleviated by confining the high effective density to the viscous regions, while computing just sufficient collisions in the isentropic regions to keep the flow in equilibrium. This may be done by setting smaller collision cross-sections for the molecules outside the viscous regions. Since the extent of the viscous regions is initially unknown, the adjustment of collision cross-sections would be best carried out interactively during the running of the program. An interactive procedure in which the distribution of inflowing molecules is set consistently with that of the outgoing molecules provides the optimum procedure for dealing with subsonic boundaries. These considerations, as well as overall cost-effectiveness, indicate that simulation studies would be best carried out with a dedicated minicomputer rather than through access to a general purpose computing centre.

The major advantage of the simulation method is that there are no stability problems. For applications in which some spreading of shock wave profiles can be tolerated, the guidelines on maximum cell size and time step that were discussed in § 7.2 may be exceeded by large factors. The only point that need be watched is that the ratio of cell size to time step should preferably be smaller than the speed of sound and should not exceed it by a large factor. Note the contrast between this and the Courant criterion for the stability of a finite difference computation. The required acoustic signal propagation mechanism is impossible in the continuum method unless the ratio of the distance to time step exceeds the local speed of sound. Acoustic signals are propagated through the intermolecular collisions in the simulation method, even though this ratio may be very small in comparison with the speed of sound. The ease with which the simulation method may be applied to problems involving curved boundaries in either two or three spatial dimensions represents a further advantage over finite difference solutions of the Navier–Stokes equations. Simulation procedures for flows involving gas mixtures, internal degrees of freedom, and chemical reactions are presented in Chapters 10, 11, and 12, respectively.

10

FLOWS INVOLVING GAS MIXTURES

10.1. General approaches

IF a gas flow is in local thermodynamic equilibrium, the question of whether it is a simple gas or a gas mixture is of no consequence. An equilibrium flow is completely described by the Euler equations in terms of the macroscopic properties that were defined for gas mixtures in § 1.3. Finite diffusion velocities and species separation effects are introduced at the level of the Navier–Stokes equations for flows with small departures from equilibrium. Even if the mixture is initially of uniform composition, separation effects may be produced by thermal and pressure diffusion. Fortunately, these can generally be ignored even when dealing with a viscous flow. When separation effects are taken into account, the diffusion velocities and species number densities are introduced as additional variables. This means that the continuity, momentum, and energy equations must be supplemented by the general diffusion equation (Chapman and Cowling 1952, p. 244) and the species continuity equations. At the other extreme, we have seen that the individual gas species in a collisionless flow move independently of one another, thus leading to very pronounced separation effects. Therefore, in a transition regime flow, the magnitude of the effects due to a gas mixture may be expected to depend on the degree to which the gas departs from thermal equilibrium.

In addition to the species separation effects and the associated finite diffusion velocities, there may be differences in the kinetic temperatures of the individual species. These may be particularly pronounced for mixtures consisting of a light gas with a small proportion of heavy gas. The frequency of heavy–heavy collisions would be small compared with that for light–light collisions. Furthermore, the cross collisions between light and heavy molecules would have little effect on the heavy gas. The effective time scale for temperature relaxation would then be quite different for the two components and this may be expected to lead to significant departures from the simple gas solutions for problems such as the internal structure of strong shock waves.

The Boltzmann equation (3.21) for a gas mixture is a series of simultaneous differential equations with the individual species distribution functions as the dependent variables. These equations are coupled through the cross collision components of the collision terms. The Chapman–Enskog solution of this equation leads to the general diffusion equation that was referred to above. As in the case of the multi-dimensional flows, the intuition that is required for successful applications of the moment method has not been forthcoming, the BGK equation presents extreme mathematical difficulties for realistic problems, and the finite difference Monte Carlo method is computationally impracticable. On the other hand, the direct simulation Monte Carlo method may be readily applied to these flows.

10.2. Direct simulation procedures

The procedures that are required for the extension of the direct simulation Monte Carlo method to gas mixtures will be described in the context of a specific application. This is the study of temperature relaxation in a homogeneous gas mixture. The initial conditions correspond to a gas consisting of an arbitrary number of species that are uniformly distributed in physical space, but which each have a different temperature. The relaxation process refers to the subsequent establishment of an equilibrium temperature as a result of the intermolecular collisions. The gas remains homogeneous in physical space and the problem may be regarded as a 'zero-dimensional' flow. A complete FORTRAN program for the direct simulation of this problem is listed in Appendix I.

The required information on the simulated molecules is stored in the three-dimensional variable P(L, M, N). Since all locations are equally probable in physical space, only velocity space information need be stored. The first subscript designates the species, while the second designates the individual molecules of that species. The inclusion of these two in a single array does not lead to any wasted storage because weighting factors are established such that there is an equal sample of molecules of each species, irrespective of the relative number densities. If there were wide differences in the number of molecules of the various species, it would be preferable to specify separate variables for each species or, alternatively, store the information in just one dimension of a single variable. In the latter case, cross reference variables, similar to the variables in the program of Appendix G, would permit easy access to the molecules of a particular species. This would be the preferred procedure for a reacting gas because a molecule could then change from one species to another without there being any necessity to transfer all the relevant information from one array to another. The third subscript in the variable P(L, M, N) ranges from 1 to 3 and distinguishes between the three velocity components u, v, and w that are stored for each molecule. Advantage could have been taken of the spherical symmetry of the distribution function in order to store only the molecular speed. The consequent reduction in storage requirements would, however, have been gained at the expense of an increase in computation time. This is because it would have been necessary to convert the molecular speeds to velocities, through the choice of a random direction, each time a molecule was considered as a possible participant in a collision.

The first of the two READ statements for the input of data sets the number of species as NC, the number of molecules of each species as NM, and sets the number NP and magnitude DTP of the time intervals between the sampling and printing of results. The second READ statement is within a loop over the NC molecular species and sets the number density, molecular mass, molecular diameter, and initial temperature of each species in turn. These quantities are stored in the two-dimensional array C(L, N).

The overall number density is, from eqn (1.28),

$$n = \sum_{p=1}^{s} n_p$$

and the overall temperature of the mixture is

$$T = \frac{1}{n} \sum_{p=1}^{s} n_p T_p. \tag{10.1}$$

In the program, the quantities p, s, n_p, n, T_p, and T are represented by the FORTRAN variables L, NC, C(L, 1) SN, C(L, 4), and TFE, respectively. In order for the time to be normalized by the mean collision time per molecule in the final equilibrium gas, the molecular diameters are assigned values such that this quantity is unity. The mean collision rate per molecule in an equilibrium gas mixture is, from eqns (4.39) to (4.41)

$$\nu_0 = \sum_{p=1}^{s} \frac{n_p}{n} \sum_{q=1}^{s} \tfrac{1}{2}\pi^{\frac{1}{2}}(d_p + d_q)^2 n_q \{2kT(m_p + m_q)/m_p m_q)\}^{\frac{1}{2}}. \tag{10.2}$$

The Boltzmann constant k is conveniently set equal to $\frac{1}{2}$ and, in order for $1/\nu_0$ to be unity, the diameter of molecule 1 is

$$d_1 = \left[\sum_{p=1}^{s} \frac{n_p}{n} \sum_{q=1}^{s} \tfrac{1}{2}\pi^{\frac{1}{2}} \left(\frac{d_p}{d_1} + \frac{d_q}{d_1} \right)^2 n_q \{T(m_p + m_q)/m_p m_q\}^{\frac{1}{2}} \right]^{-\frac{1}{2}}.$$

In the program, the diameters were altered to ratios within the loop over label 8 and the summation is carried out in the loops over label 9. The required diameter of molecular species 1 is then set as C(1, 3), with the other molecular diameters being set in the loop over label 10. The collision rates are to be sampled and compared with theory. The theoretical collision rates for a molecule of species L with one of species M is calculated as TC(L, M) through a direct application of eqn (4.39) in the loops over label 36. These loops are also used to set the time parameter T(L, M) and collision counter SC(L, M) to zero, and to set the maximum probable relative velocity in a collision VM(L, M) to a reasonable initial value.

Had the more general inverse-power law molecular model been used in place of the hard-sphere model, the normalization of the results would have presented a far more difficult problem. The conventional solution to the problem posed by the indeterminate diameter of an inverse power law molecule is to combine the Chapman–Enskog viscosity coefficient result with the hard-sphere relationship between this coefficient of viscosity and the mean free path. The details of this procedure were presented for a simple gas in § 8.1. In the case of gas mixtures, the Chapman–Enskog coefficient of viscosity is not readily available for the inverse power low model. However, this model will generally be used only when the simulation is intended to represent a mixture of real gases and the normalization can then be dispensed with. The effective viscosity diameters (see Appendix A) may then be used for each molecular species, with all other variables and constants given their actual numerical values.

The initial values of the molecular velocity components are set in the loops over the label 12. It is assumed each gas species is initially in equilibrium at the temperature stored in C(L, N), thus enabling the most probable molecular speed for each species (with $k = \frac{1}{2}$) to be calculated and stored as D. The typical velocity components are then generated by the acceptance–rejection method, following the example in Appendix D. This completes the preliminaries; the next step is to initiate the main loop over the NP time intervals.

The flow time is stored as the FORTRAN variable TIME and this is advanced by the time interval DTP as the first statement in the main loop. The number of 'classes' of collision is equal to the square of the number of molecular species and we will refer to 'class L-M', where both L and M range from 1 to NC. These are computed in the two loops over label 15; sufficient collisions being computed to bring the time counter T(L, M) for each class up to the flow time. The first step in the collision routine is the random selection of molecule K of species L and molecule J of species M. The three components of the relative velocity between this pair are then calculated as VRC(N), and the magnitude of the relative velocity as VR. Since the probability of the randomly selected pair participating in a collision is proportional to VR, the pair is accepted or rejected on this basis. The probability VR is normalised to a maximum probability of unity by dividing it by VM(L, M), but with the maximum probable relative velocity VM(L, M) being increased to VR if the previous estimate proves to be too low.

Once the collision pair has been selected, account must be taken of the weighting factors of the two molecules. All species are represented by the same number of simulated molecules, even though the number densities are unequal. Each simulated molecule of a gas species with a low number density represents proportionally fewer actual molecules than a simulated molecule of a more abundant species. The species weighting factor is, therefore, equal to the number density of that species; this is set as C(L, 5) in the loop over label 8. When a collision occurs between molecules with different weighting factors, the velocity of the molecule with the smaller factor is always changed in the collision, while that for molecule with the larger factor is changed only with probability proportional to the ratio of the smaller to the larger factor. In the program, the variables LP and MP represent the probabilities that the collision will be effective for the L and M molecules, respectively. These probabilities are initially set to unity; that of the molecule with the larges weighting factor is then either allowed to remain unity or is set to zero through an acceptance–rejection comparison based on the ratio of weighting factors.

The generalization of eqn (7.4) for the time increment Δt_c is, for a type L–M collision,

$$\Delta t_c = \frac{\mathrm{LP}}{N_L}\frac{1}{\sigma_{LM} n_M c_r} + \frac{\mathrm{MP}}{N_M}\frac{1}{\sigma_{LM} n_L c_r}. \tag{10.3}$$

The cross section σ_{LM} is equal to $\pi(d_L + d_M)^2/4$ and is denoted by CXS. In this application both N_L and N_M are represented by NM. The first and second terms in eqn (10.3) account for the contributions to the time increment due to the L and M molecules, respectively. The first term is therefore multiplied by the probability LP, and the second by the probability MP. The advancing of the counter SC for the collisions per molecule also depends on these probabilities. New components of the relative velocity vector are calculated by the same routine that was described for the simple gas in § 7.3 and listed in Appendix G. The centre of mass velocity is calculated from eqn (2.1) and the new velocity components follow from eqn (2.5).

The final routines are for the sampling and printing of the collision rates and temperatures. The average collision rate for the mixture as a whole is calculated from eqn (1.32) in the loops over label 25. The overall temperature is calculated from eqn (1.41) and the individual species temperatures from eqn (1.42). The overall temperature should, of course, be constant throughout the calculation. However, the use of weighting factors means that both molecular velocities are not necessarily altered in a collision so that exact energy conservation is not imposed by the equations. The level of fluctuation in the overall temperature provides a measure of this undesirable side effect that is unavoidably introduced by the use of weighting factors.

10.3. Typical flows

We will first consider some results from the homogeneous gas relaxation program that was described in the preceding section. The test case of Appendix

I deals with a three-component mixture such that

$$n_1:n_2:n_3 = 1:1:1,$$

$$m_1:m_2:m_3 = 1:3:10,$$

and

$$d_1:d_2:d_3 = 1:2:0{\cdot}5.$$

The subscripts 1, 2, and 3 indicate the three species; the initial temperatures of these species being 9, 5, and 1, respectively. The overall temperature is given by eqn (10.1) as 5·0, while the normalization of the program is such that the equilibrium collision frequency is unity. The theoretical equilibrium collision rate for all combinations of collision pairs are also listed at the beginning of the output. The first and last pages of the output for a run with a sample of 4000 per species are reproduced at the end of Appendix I. This follows the relaxation process in the gas to a normalized time of 30·0. As shown in Fig. 10.1, the three components of the gas come to equilibrium at approximately twenty mean collision times. The sampled collision rates

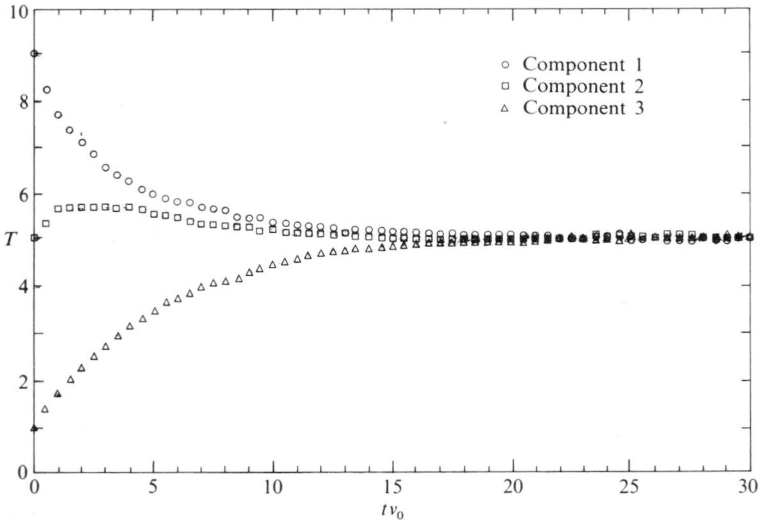

FIG. 10.1. Temperature relaxation in a homogeneous three component gas mixture.

in the equilibrium state are in agreement with theoretical values. All collision rates for pairs which include a species three molecule are smaller than the average rate. For this reason, the molecules of species one and two tend to come to equilibrium with one another before either of them comes to equilibrium with species three. This causes the temperature of the species two gas to increase to a maximum before returning to its original value which coincides with the equilibrium temperature. This illustrates the point that a number of distinct time scales may be associated with multiple gas mixtures.

The overall temperature in the above example was 5·0135 at all times. The departure from the theoretical value of 5 is a consequence of the statistical scatter associated with the initial generation of typical molecular velocities. A similar calculation was carried out with the number densities in the ratio of

$$n_1:n_2:n_3 = 10:2:1.$$

The sample size was again 4000 per species and the number density variations were established through weighting factors. The relaxation process was generally similar to those in the above example, but the overall temperature was dominated by that of species 1 and the initial departure from the theoretical value was almost one per cent. Also, the use of weighting factors led to a subsequent fluctuation of this temperature, as explained in the preceding section. The level of the fluctuation in this example was approximately two per cent.

The simulation routines for dealing with gas mixtures have been incorporated in the program for the study of the internal structure of shock waves. The simulation routines were described in the preceding section and the shock structure program was outlined in § 8.1. A shock wave of Mach number 10 in binary mixture of hard-sphere molecules has been chosen as a demonstration case. The mass ratio is 10:1, the undisturbed number density ratio is 1:10, and both molecules have the same diameter. The normalized profile of the steady shock wave is shown in Fig. 10.2. The light-gas density profile leads the heavy-gas profile by a distance that is approximately equal to five mean free paths in the undisturbed upstream gas. The density profile of the mixture lies between the profiles of the individual species. The lag in the heavy gas profile means that there is a local concentration of light gas within the shock wave. When the shock wave is generated by a piston moving into the gas, as in the program under consideration, this must be balanced by a concentration of heavy gas adjacent to the piston.

As noted in § 8.1, the simulation program may be used to study unsteady shock wave formation in addition to the steady wave profile. This capability is illustrated in Fig. 10.3 which shows the concentration ratio profile ahead of the piston at a number of time instants. The normalized concentration ratio $\tilde{n}$ is obtained by dividing the ratio of the light gas to heavy gas number density by the undisturbed number density ratio. For small times, the concentration ratio profile resembles the collisionless profile that is readily obtained from the theory of § 5.3. The fully formed shock wave profile does not separate from the piston concentration profile until a time of approximately $4\lambda_1/a_1$. The profile of the concentration ratio adjacent to the piston

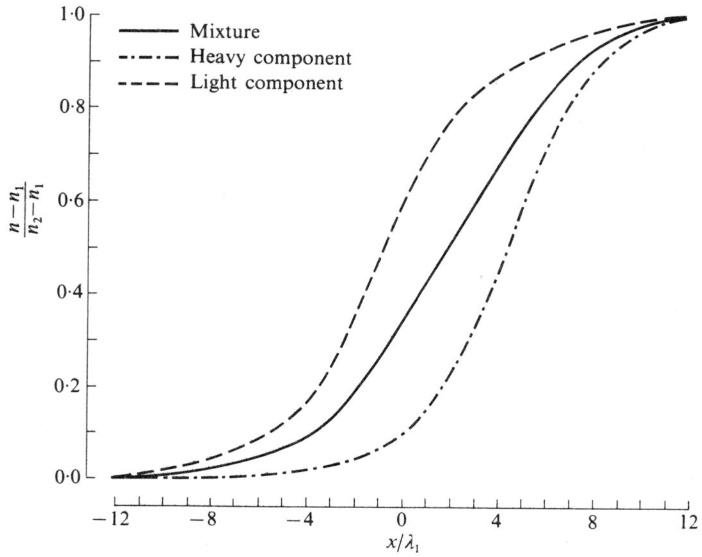

FIG. 10.2. Number density profiles for $(Ma)_s = 10$ shock wave in a binary gas mixture.

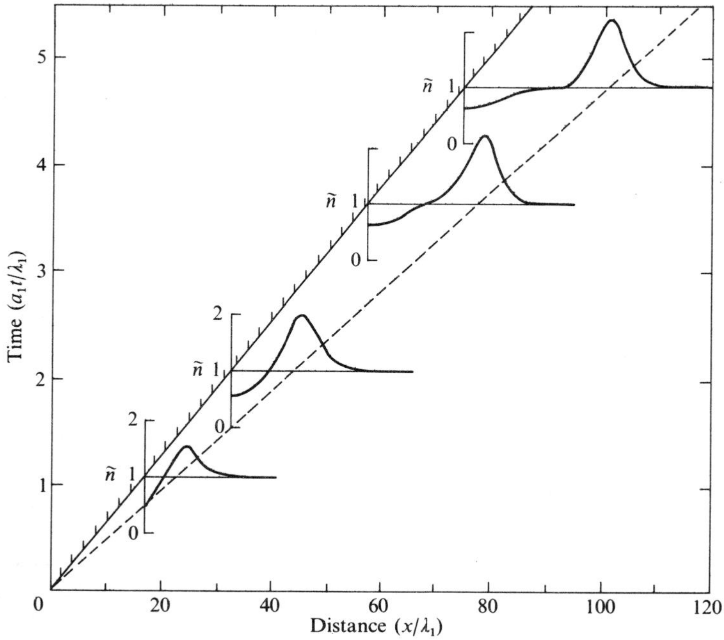

FIG. 10.3. Contours of constant concentration ratio during the formation of a shock wave of $(Ma)_s = 10$ in a binary gas mixture.

does not attain a steady form and gradually diffuses away from the piston as its magnitude decreases.

The simulation program for two-dimensional and axially symmetric flows that was described in § 9.2 has also been extended to deal with gas mixtures. Fig. 10.4 shows contours of constant concentration ratio in

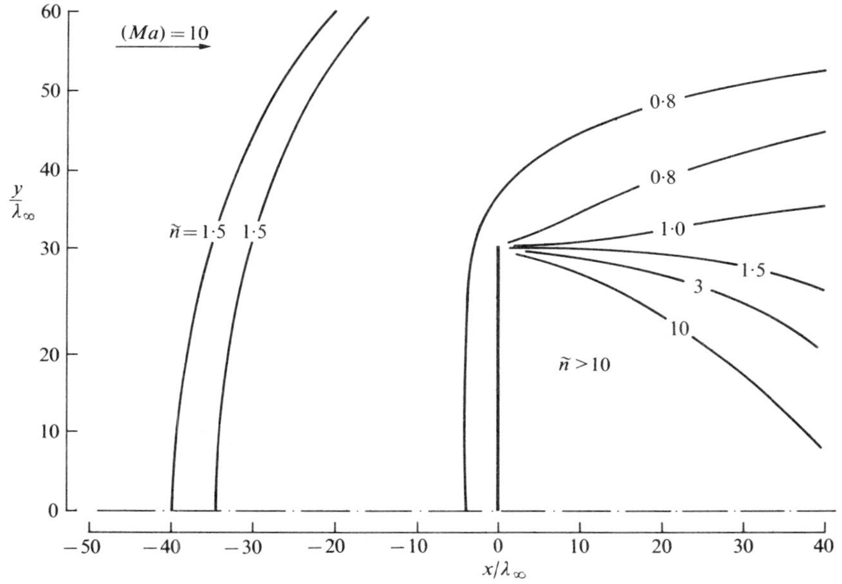

Fig. 10.4. Contours of constant concentration ratio in the steady flow of a binary gas mixture past a vertical flat plate.

the two-dimensional steady flow of a binary gas past a vertical flat plate. The undisturbed composition of the gas is identical with that in the previous example of one-dimensional shock formation and the concentration ratio $\tilde{n}$ is defined in the same way. The shock wave structure near the centre line is similar to that in the one-dimensional problem and is marked by a local increase in the light gas concentration. There is an increase in the heavy gas concentration adjacent to the front face of the plate. This is analogous to the concentration of heavy gas adjacent to the piston in the shock formation problem. The most pronounced separation effect occurs downstream of the plate where there is a large region composed almost entirely of light gas. The average density in the region for which the concentration ratio exceeds ten is approximately one twentieth of the freestream number density. This region of greatly enhanced light gas concentration downstream of the body is bounded by a region in which the light gas is depleted.

11

DIATOMIC AND POLYATOMIC GAS FLOWS

11.1. Models with internal degrees of freedom

THE analysis of gas flows involving molecules with internal energy must take account of the interchange of internal and translational energy during collisions. The existence of internal energy was taken into account in the discussion of the molecular-gas model in Chapter 1 and in the analysis of collisionless flows in Chapter 5. However, in the analysis of collision mechanics in Chapter 2 and in all subsequent analysis that involves intermolecular collisions, consideration has been restricted to molecular models without internal energy. The internal energy may be contained in both rotational and vibrational degrees of freedom. While classical theory cannot be expected to lead to an adequate treatment of any active vibrational modes, it should be adequate for the rotational modes. However, the difficulties associated with conventional molecular models that involve rotation have greatly hampered the application of kinetic theory to diatomic and polyatomic gas flows.

For monatomic gases, the inverse power law model generally provides the optimum balance between ease of analysis and accuracy of representation. The corresponding model for a diatomic gas would be two power law centres of repulsion separated by a fixed distance along the internuclear axis. This model has been studied by Lordi and Mates (1970); a complex numerical solution being required for each set of impact parameters. It appears to be quite out of the question to use this model for Monte Carlo simulations that require the computation of many thousands, or even millions, of collisions.

The best prospects for a workable polyatomic model would appear to lie with some modification of the hard sphere model with its very simple collision mechanics. The *sphero-cylinder molecule*, consisting of a smooth cylinder with hemispherical ends, is a geometrical variation with apparent relevance to diatomic gases. The collision mechanics of sphero-cylinders have been developed by Curtiss and Muckenfuss (1958). The model requires additional variables for the specification of the spatial orientation, as well as for the angular velocity. A particular difficulty is that this orientation, and therefore the collision cross-section, changes with time. Whether or not two molecules collide depends not only on their miss distance, but also on whether or not they 'mesh' or 'clash'. Moreover, there may be multiple 'chattering' impacts in a single collision event. The *weighted-sphere model* had earlier been suggested by Jeans (1904) and has been extensively developed by Dahler and Sather (1962) and Sandler and Dahler (1967). Although it is spherical in geometry, the molecule rotates about the offset centre of gravity rather than the geometrical centre. It therefore suffers from essentially the same disadvantages as the sphero-cylinder model. The only rotating model with a cross-section that is not affected by its orientation is the *rough-sphere model.* The name of this model is based on its basic physical property which is that the velocities of the points of contact are reversed in a collision. This model was first suggested by Bryan (1894) and the transport properties of a rough-sphere gas were analysed by Pidduck (1922).

Melville (1972) found that a modified weighted-sphere model is the most complicated one that can be handled in a direct simulation Monte Carlo program. The modification involved the neglect of the 'wobbling' about the centre of mass. Even so, the computer storage and time requirements were significantly greater than those for monatomic models. Moreover, the weighted-sphere model is restricted to 'hard' interactions with very limited control over the effective relaxation time. The rough-sphere model may be applied exactly and its collision mechanics are described in detail in § 11.2. One drawback is that the model has three rotational degrees of freedom, and therefore has a specific heat ratio of 4/3, rather than the generally required value of 7/5 for a diatomic gas. There are further serious objections to the artificiality of the rough-sphere model (Chapman and Cowling 1970, § 11.62).

The limitations of the models based on exact dynamical systems are sufficient to justify recourse to a phenomenological approach. The basic requirements for a satisfactory model for simulation studies are:

(*i*) The computation time and storage requirements should be comparable with those for the monatomic gas models.

(*ii*) The model should contain a specified number of internal degrees of freedom such that, in equilibrium, there is equipartition of energy between them.

(*iii*) The model should not contain more adjustable parameters than are required to specify the 'hardness' of the model and the relaxation times.

Several models have been put forward in response to these requirements and these are described and discussed in § 11.3.

The phenomenological models do not distinguish between the modes of internal energy; the number of internal degrees of freedom being a fully adjustable parameter. The energy attributable to the rotational and vibrational modes could be distinguished and separate relaxation times assigned to them. The use of a phenomenological model does, however, imply an averaging process that is generally appropriate only to fully excited modes containing a very large number of quantum states. For gases such as oxygen and nitrogen, significant excitation of the vibrational modes commences at temperatures of the order of 1000 K. The phenomenological model would then require the specification of a temperature or velocity dependent number of internal modes. However, a preferable approach is based on the fact that the number of populated vibrational states is generally sufficiently small to allow each state to be regarded as a separate species in a gas mixture. The simulation procedures for gas mixtures that were discussed in Chapter 10 may be readily extended to cope with collisional transitions from one vibrational state to another. Information is available on many of the cross-sections for the vibrational transitions and the ratios of these to the elastic cross-sections give the probabilities of transition at each collision. Many of the transitions involve the emission of photons and radiative effects may also be incorporated into the simulation.

11.2. The rough-sphere molecular model

The essential feature of the rough sphere model is that, in a collision, the relative velocity at the point of contact is reversed. Both the translational and rotational velocities contribute to this relative velocity and its reversal generally leads to an interchange of energy between the rotational and translational modes, as well as between the collision partners. The total energy is, of course, conserved. The collision mechanics will be presented in a form which allows a straightforward incorporation of the model into the direct simulation Monte Carlo method. The rough sphere is the simplest of the real physical models and flow solutions which employ it may be used as standards against which the corresponding solutions employing the phenomenological models may be judged. Since this may be done without recourse to gas mixtures, the analysis will be limited to the simple gas case.

The angular velocity of a molecule is denoted by $\boldsymbol{\omega}$ and its components about the x, y, and z axes are ω_x, ω_y, and ω_z, respectively. The rotational energy of a molecule is $\frac{1}{2}I\omega^2$, where I is its moment of inertia. The moment of inertia depends upon the radial mass distribution of the molecule and therefore allows a limited range of variability in the model. It is most convenient to take a uniform mass distribution for which

$$I = md^2/10. \tag{11.1}$$

The combined normalized distribution function is (Chapman and Cowling 1952)

$$f = \frac{(mI)^{\frac{3}{2}}}{(2\pi kT)^3} \exp\{-(mc'^2 + I\omega^2)/(2kT)\}$$

and this may be written

$$f = (\beta^3/\pi^{\frac{3}{2}}) \exp(-\beta^2 c'^2)\{1/(\pi^{\frac{3}{2}}\omega_m^{\frac{3}{2}})\} \exp(-\omega^2/\omega_m^2). \tag{11.2}$$

Here, ω_m is the most probable angular speed and is related to the temperature by

$$\omega_m = (2kT/I)^{\frac{1}{2}}, \tag{11.3}$$

or, using eqn (11.1) for the uniform mass distribution,

$$\omega_m = (20\,RT)^{\frac{1}{2}}/d. \tag{11.4}$$

The equilibrium distribution of the angular speed ω with respect to the most probable value is, therefore, completely analogous to the distribution of the thermal speed with respect to the most probable molecular speed. All directions are equally possible for the axis of rotation. A rotational temperature may be defined by

$$\tfrac{3}{2}kT_{\mathrm{rot}} = \tfrac{1}{2}I\overline{\omega^2}. \tag{11.5}$$

The precollision values of the translational and rotational velocities of a collision pair are $\boldsymbol{c}_1$, $\boldsymbol{c}_2$, and $\boldsymbol{\omega}_1$, $\boldsymbol{\omega}_2$ respectively. These are regarded as known quantities and the objective is to determine the post-collision values $\boldsymbol{c}_1^*$, $\boldsymbol{c}_2^*$, $\boldsymbol{\omega}_1^*$, and $\boldsymbol{\omega}_2^*$.

The pre-collision velocity vector $\boldsymbol{c}_{\mathrm{r}}$ is defined for this section by $\boldsymbol{c}_{\mathrm{r}} = \boldsymbol{c}_2 - \boldsymbol{c}_1$ (note the opposite sign convention to that eqn (2.3)). Typical impact parameters b and ε may be chosen. The parameter b is distributed between 0 and d with probability proportional to b, while ε is uniformly distributed between 0 and 2π. It is convenient to define three vectors $\boldsymbol{j}$, $\boldsymbol{k}$, and $\boldsymbol{l}$ in the collision plane, as shown in Fig. 11.1. The vector $\boldsymbol{k}$ is defined by the line

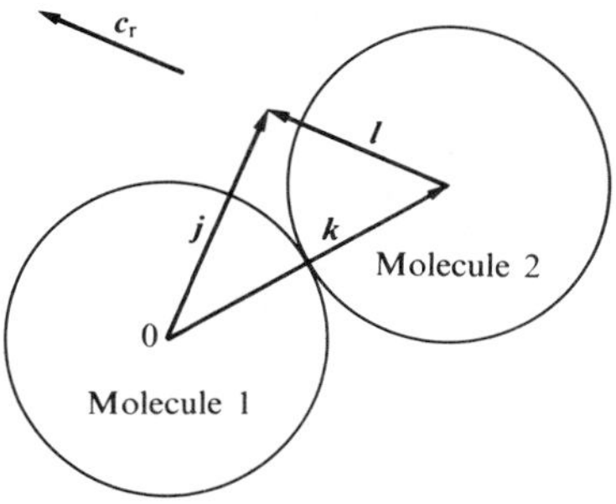

FIG. 11.1 Illustration of the vectors, $\boldsymbol{j}$, $\boldsymbol{k}$, and $\boldsymbol{l}$ in a binary collision of rough-sphere molecules.

joining the centre of molecule 1, which is taken as the origin 0, to the centre of molecule 2. The vector $\boldsymbol{l}$ is in the direction of the relative velocity vector $\boldsymbol{c}_{\mathrm{r}}$ and joins the centre of molecule 2 to the plane through 0 normal to the direction of $\boldsymbol{c}_{\mathrm{r}}$ and $\boldsymbol{l}$. Finally the impact vector $\boldsymbol{j}$ is defined by $\boldsymbol{j} = \boldsymbol{k} + \boldsymbol{l}$ and its magnitude is equal to the impact parameter b. Therefore, since the magnitude of $\boldsymbol{k}$ is equal to d and $\boldsymbol{l}$ is normal to $\boldsymbol{j}$,

$$l = (d^2 - b^2)^{\frac{1}{2}} \tag{11.6}$$

Angles θ and ϕ are now defined such that θ is the angle between the x axis and $\boldsymbol{c}_r$, and ϕ is the angle between the y axis and the plane containing the x axis and $\boldsymbol{c}_r$. Then

$$\cos\theta = u_r/c_r$$

and (11.7)

$$\cos\phi = v_r/(c_r \sin\theta).$$

Then, with the azimuth impact parameter measured from the plane $y = 0$, the components of the impact vector are

$$j_x = b\cos\varepsilon\sin\theta,$$

$$j_y = -b(\cos\varepsilon\cos\theta\cos\phi + \sin\varepsilon\sin\phi),$$

and (11.8)

$$j_z = b(\sin\varepsilon\cos\phi - \cos\varepsilon\cos\theta\sin\phi).$$

The magnitude of $\boldsymbol{l}$ is given by eqn (11.6) and its components are

$$l_x = lu_r/c_r,$$

$$l_y = lv_r/c_r, \tag{11.9}$$

and

$$l_z = lw_r/c_r.$$

The components k_x, k_y, and k_z of the vector $\boldsymbol{k}$ along the line of centres follow from the relation $\boldsymbol{k} = \boldsymbol{j} - \boldsymbol{l}$. The components of the relative velocity of the points of contact are then given by

$$g_x = u_r + \tfrac{1}{2}\{k_y(\omega_{z_1} + \omega_{z_2}) - k_z(\omega_{y_1} + \omega_{y_2})\},$$

$$g_y = v_r + \tfrac{1}{2}\{k_z(\omega_{x_1} + \omega_{x_2}) - k_x(\omega_{z_1} + \omega_{z_2})\}, \tag{11.10}$$

and

$$g_z = w_r + \tfrac{1}{2}\{k_x(\omega_{y_1} + \omega_{y_2}) - k_y(\omega_{x_1} + \omega_{x_2})\}.$$

The scalar product of $\boldsymbol{k}$ and $\boldsymbol{c}_r$ is denoted by k_c, so that

$$k_c = \boldsymbol{k} \cdot \boldsymbol{c}_r = k_x u_r + k_y v_r + k_z w_r.$$

The post-collision values of the translational velocity components may now be related to the pre-collision values by

$$u_1^* - u_1 = u_2 - u_2^* = \tfrac{2}{7}g_x + \tfrac{5}{7}k_c k_x/d^2,$$

$$v_1^* - v_1 = v_2 - v_2^* = \tfrac{2}{7}g_y + \tfrac{5}{7}k_c k_y/d^2, \tag{11.11}$$

and

$$w_1^* - w_1 = w_2 - w_2^* = \tfrac{2}{7}g_z + \tfrac{5}{7}k_c k_z/d^2.$$

Finally, the post-collision angular velocity components are given by

$$\omega_{x_1}^* - \omega_{x_1} = \omega_{x_2}^* - \omega_{x_2} = \tfrac{10}{7}(k_y g_z - k_z g_y)/d^2,$$

$$\omega_{y_1}^* - \omega_{y_1} = \omega_{y_2}^* - \omega_{y_2} = \tfrac{10}{7}(k_z g_x - k_x g_z)/d^2, \tag{11.12}$$

and

$$\omega_{z_1}^* - \omega_{z_1} = \omega_{z_2}^* - \omega_{z_2} = \tfrac{10}{7}(k_x g_y - d_y g_x)/d^2.$$

11.3. Phenomenological models

The first model to be considered is the 'energy sink' model that was put forward by Bird (1970*c*). In this, a single additional variable representing the internal energy me_{int} (or ε_{int}) is associated with each molecule. This internal energy is generally the total internal energy summed over all degrees of freedom, although it is possible to introduce separate variables for fully excited rotational and vibrational modes which may require different relaxation times. The special case of partially excited vibrational states is better dealt by treating each state as a separate chemical species, as noted in § 11.2. The energy-sink model is essentially a modification of the inverse power law model, with the hard sphere and Maxwell molecules as special cases; it remains spherically symmetric.

The average internal energy of a molecule with ζ internal degrees of freedom is, from eqn (1.21) and assuming equipartition,

$$m\bar{e}_{\text{int}} = (\zeta/6)\overline{mc'^2}. \tag{11.13}$$

The average relative translational energy between collision partners in an equilibrium gas of inverse power law molecules is, from eqn (4.45),

$$\tfrac{1}{2}\overline{m_r c_r^2} = \{2(\eta-2)/(\eta-1)\}kT. \tag{11.14}$$

Therefore, since $\tfrac{3}{2}kT = \tfrac{1}{2}\overline{mc'^2}$, the average internal energy of a molecule is related to the average relative translational energy in collisions in an equilibrium gas by

$$m\bar{e}_{\text{int}} = (\zeta/8)\{(\eta-1)/(\eta-2)\}\overline{m_r c_r^2}. \tag{11.15}$$

This equation is the basis of the model, in that the value of me_{int} for each collision partner is compared with the value of $(\zeta/8)\{(\eta-1)/(\eta-2)\}m_r c_r^2$ for the collision. If me_{int} is not equal to $(\zeta/8)\{(\eta-1)/(\eta-2)\}m_r c_r^2$, a fraction t_f of the difference is transferred between the translational and internal modes in the direction to more nearly satisfy the equilibrium eqn (11.15). The net change in the translational energy is compensated by an appropriate decrease or increase in the magnitude c_r^* of the post-collision relative velocity. This modifies the post-collision translational velocities which are calculated from $\boldsymbol{c}_r^*$ and the unchanged $\boldsymbol{c}_m$ by eqn (2.5).

The details of the model are best illustrated through a demonstration program for the relaxation of the internal degrees of freedom in a homogeneous gas. The program is listed in Appendix J and makes provision for both the hard sphere and inverse ninth power molecular models. The three velocity components and the internal energy are stored for each molecule in the two-dimensional array P(4, M). The only other array required for this program is that for the components of the relative velocity in a collision. The data consists of the total number of simulated molecules, a code number to indicate whether hard sphere or inverse ninth power collision mechanics are to be used, the number of internal degrees of freedom, the transfer factor t_f, the initial ratio of the internal to the translational kinetic temperature, the intervals at which the temperatures are to be sampled, and the total number of intervals in a run. The time is normalized to the product of the mean free path and the inverse of the molecular most probable molecular speed at the initial translational temperature. The mean collision time would have been more appropriate for this normalization if the program had been confined to the hard sphere model.

The initial density, translational temperature, most probable equilibrium molecular speed, mean free path, and molecular mass are conveniently set equal to unity. This means that both the gas constant and the Boltzmann constant are numerically equal to one half. The collision cross-section of a hard sphere molecule is then given by eqn (4.38) as $1/\sqrt{2}$ and is set as CXS. The collision cross-section for the inverse ninth power molecules is dependent on the relative velocity and, for this model, the constant parameter CXS is most conveniently set equal to the expression given by eqn (8.17). This enables it to be used in a similar manner for each model in the evaluation of eqn (7.2) for the time interval to be attributed to a collision. The parameter BN is set equal to the coefficient of kT in eqn (11.14) for the average translational energy in a collision in an equilibrium gas. The initial translational velocity components of the molecules are set by the same acceptance–rejection routine that has been used in the demonstration program of Appendix I. Since the initial translational temperature is unity, the initial internal temperature is given by the parameter TIN which is set in the data as the initial ratio of the internal to the translational temperature. The internal energy of each molecule is set equal to the equilibrium mean value of $\frac{1}{2}kT$ per degree of freedom.

The collision routine commences with the random selection of a pair of molecules with subscripts J and L. The relative velocity components are set in the array VRC(3), with the magnitude and the square of the magnitude of the relative velocity being set as VR and VRR, respectively. The parameter VRE is set equal to VR for the hard-sphere molecules and the square root of VR for the inverse ninth power law of molecules. The normalized collision probability for the selected pair of molecules is then equal to VRE/VRM, where VRM is the maximum likely value of either the relative speed or its square root. This had been set as twice the most probable molecular speed for the hard sphere case and one and a half times this speed for the inverse ninth power case. Should VRE exceed VRM, the variable VRM is reset so that the normalized probability cannot exceed unity. A simple acceptance–rejection procedure is then used to either accept or reject the pair with appropriate probability.

For the pairs that are accepted for a collision, the time counter is advanced through the interval given by eqn (7.2) and the collision counter is advanced by unity. The working variable A is then set equal to the value of $\frac{1}{8}\zeta\{(\eta-1)/(\eta-2)\}m_r c_r^2$ for this collision. The change in the internal energy of each molecule is equal to the product of the transfer factor TFAC and the difference between the internal energy and A. This is set equal to DERJ for molecule J and DERL for molecule L, and the internal energies P(4, J) and P(4, L) are adjusted accordingly. The translational kinetic energy is $\frac{1}{2}m_r v_r^2$ or, since m_r is here equal to $\frac{1}{2}$, VRR/4. Therefore, in order to add the sum of DERL and DERJ to the translational energy, VRR is increased by four times this sum. The parameter VRF is then set equal to the ratio of the new to the old relative speed.

The next step is to calculate the post-collision components of the relative velocity and, for this, a separate routine is required for each model. The inverse ninth power routine is dealt with first. The collision mechanics of this model were analysed in Chapter 2 and its implementation in a direct simulation program was outlined in § 8.1. The dimensionless impact parameter W_0 is set as WA in the first statement of the routine through a direct application of eqn (8.18). Eqn (8.20) is used to set the other impact parameter ε as EPS and the deflection angle χ is set as CHI through eqn (8.19). The three post-collision components of the relative velocity vector are then set through a direct application of eqn (2.22). These are set as the variable VRC(N) with the subscripts $N = 1$, 2, and 3 denoting the x, y, and z components, respectively. Note that the energy-transfer factor VRF is introduced as a scaling factor in the final steps. The hard-sphere routine follows the standard procedures and commences at the statement designated by label 17. The setting of post-collision velocity components for each molecule is common to both models and follows the standard procedures used in the earlier programs.

The final routine is for the sampling and printing of the translational and internal temperatures at intervals DTP. This commences at the statement labelled 13, control having been transferred from that labelled 12. The translational temperature TTR, the internal temperature TROT, and the overall temperature TOV are sampled through direct applications of eqns (1.23), (1.25), and (1.26), respectively. Some typical results from this program are discussed in § 11.4.

A quite distinct phenomenological model has been proposed by Larsen and Borgnakke (1973 and 1974). In this, both the relative translational and the internal energies are assumed to be distributed according to their respective equilibrium distributions and, at each collision, new values are sampled at random from these distributions, subject to the condition that the total energy is conserved. In the original form of the method, the relaxation time is adjusted by setting varying proportions of elastic and inelastic collisions. An alternative method treats all collisions as inelastic, but restricts the proportion of the energy of the collision pair that can participate in the energy exchange. We will now outline the procedure for effecting the energy transfer in a completely inelastic collision.

The distribution function for the relative speed c_r between random pairs of molecules is given by eqn (4.33) and it was shown in § 4.3 that the probability of a collision is proportional to $c_r^{(\eta-5)/(\eta-1)}$. The relative translational energy in a collision is $E_t = \frac{1}{2}m_r c_r^2$ and, noting that $\mathrm{d}E_t = m_r c_r \,\mathrm{d}c_r$, these results for the distribution of c_r lead to the following result for the distribution of E_t in collisions;

$$f_{E_t} \propto E_t^{(\eta-3)/(\eta-1)} \exp(-E_t/kT). \tag{11.16}$$

The internal energy E_i of a collision pair is the sum of the internal energies of the components i.e.

$$E_i = \varepsilon_{\mathrm{int},1} + \varepsilon_{\mathrm{int},2}. \tag{11.17}$$

The distribution function for the internal energy of a single molecule may be written (Hinshelwood 1940),

$$f_{\varepsilon_{\mathrm{int}}} \propto \varepsilon_{\mathrm{int}}^{\zeta/2-1} \exp(-\varepsilon_{\mathrm{int}}/kT). \tag{11.18}$$

We now require the distribution function f_{E_i} for E_i. Consider the fraction of collision pairs with a particular value $\varepsilon_{\mathrm{int},1}$ of internal energy in molecule 1, and therefore with $E_i - \varepsilon_{\mathrm{int},1}$ in molecule 2. Using eqns (11.17) and (11.18), and noting that $\mathrm{d}E_i = \mathrm{d}\varepsilon_{\mathrm{int},2}$ for fixed $\varepsilon_{\mathrm{int},1}$, this fraction is proportional to

$$(\varepsilon_{\mathrm{int},1})^{\zeta(2-1)}(E_i - \varepsilon_{\mathrm{int},1})^{\zeta/2-1} \exp(-E_i/kT)\,\mathrm{d}\varepsilon_{\mathrm{int},1}\,\mathrm{d}E_i. \tag{11.19}$$

The total fraction of pairs with internal energy E_i is proportional to the expression obtained by integrating eqn (11.19) over all $\varepsilon_{\text{int},1}$, from 0 to E_i. Therefore,

$$f_{E_i} \propto E_i^{\zeta-1} \exp(-E_i/kT). \tag{11.20}$$

The total energy E_c in the collision is the sum of the relative translation energy and the combined internal energy i.e.

$$E_c = E_t + E_i. \tag{11.21}$$

The probability of a particular pair of values of E_t and E_i is proportional product of f_{E_t} and f_{E_i} i.e.

$$E_t^{(\eta-3)/(\eta-1)} E_i^{\zeta-1} \exp\{-(E_t + E_i)/kT\}$$

or, using eqn (11.21),

$$E_t^{(\eta-3)/(\eta-1)} (E_c - E_t)^{\zeta-1} \exp(-E_c/kT).$$

The effective temperature T is defined by the total energy E_c in the collision, so that the exponential term may be regarded as a constant. The distribution function for the translational energy is therefore proportional to

$$E_t^{(\eta-3)/(\eta-1)} (E_c - E_t)^{\zeta-1}. \tag{11.22}$$

At each collision, a post collision value E_t^* is sampled from this distribution by the acceptance–rejection method. Note that this does not require knowledge of the constant of proportionality. The total energy in the collision is conserved and the value of the pair internal energy is, from eqn (11.21),

$$E_i^* = E_c - E_t^*. \tag{11.23}$$

Eqn (11.19) shows that the distribution function for the assignment of part of this (fixed) energy to a particular molecule is proportional to

$$(\varepsilon_{\text{int}})^{\zeta/2-1} (E_i^* - \varepsilon_{\text{int}})^{\zeta/2-1}, \tag{11.24}$$

and the acceptance–rejection method is again applied at this step. The calculation of the individual velocity components is similar to that for the energy sink model.

The advantage of Larsen's method is that the principle of detailed balance is satisfied for the translational velocities and the equilibrium distribution function must conform to the Maxwellian distribution.

11.4. Typical applications

We will consider some typical flows and compare experimental results with calculations employing the rough-sphere and phenomenological models. The major objective is to determine the extent to which the flexible and easily applied models provide an adequate representation of a real gas. The first step is to check whether the phenomenological models follow the accepted rotational relaxation equation when applied to a homogeneous gas.

Results for rotational relaxation are generally fitted to an equation of the form

$$\frac{dT_{int}}{dt} = \frac{1}{t_r}(T_e - T_{int})$$

or

$$T_{int} = T_e - (T_e - T_{int,0}) \exp(-t/t_r) \tag{11.25}$$

Here, t_r is the relaxation time, T_e is the equilibrium temperature, and $T_{int,0}$ is the value of T_{int} at time $t = 0$. Fig. 11.2 shows the time history of the rotational and translational temperatures in a homogeneous gas with two internal degrees of freedom, but with the energy initially confined to the translational mode. The calculation was made with the demonstration program that has been described in the previous section, and the first few pages of the output follow the listing of the program in Appendix J. The transfer factor t_f was equal to 0·1 for the calculation and this led to a relaxation time of approximately 6·7 mean collision times. Further calculations showed that the relaxation time is independent of the initial degree of non-equilibrium and is inversely proportional to the transfer factor.

Similar relaxation calculations to that shown in Fig. 11.2 were made for rough sphere molecules. It was found that the hard-sphere energy-sink

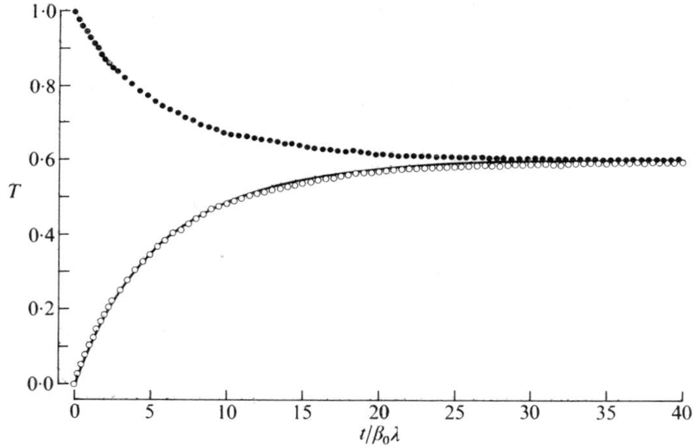

FIG. 11.2. Translational and internal temperature relaxation in a homogeneous gas with $\gamma = 7/5$. —— T_{int} from eqn (11.25) with $t_r = 6{\cdot}7\tau_0 = 5{\cdot}398\beta_0\lambda$; ● and ○ T_{tr} and T_{int}, respectively, from a direct simulation calculation employing the energy sink model with $\zeta = 2$, $\eta = \infty$, and $t_f = 0{\cdot}1$.

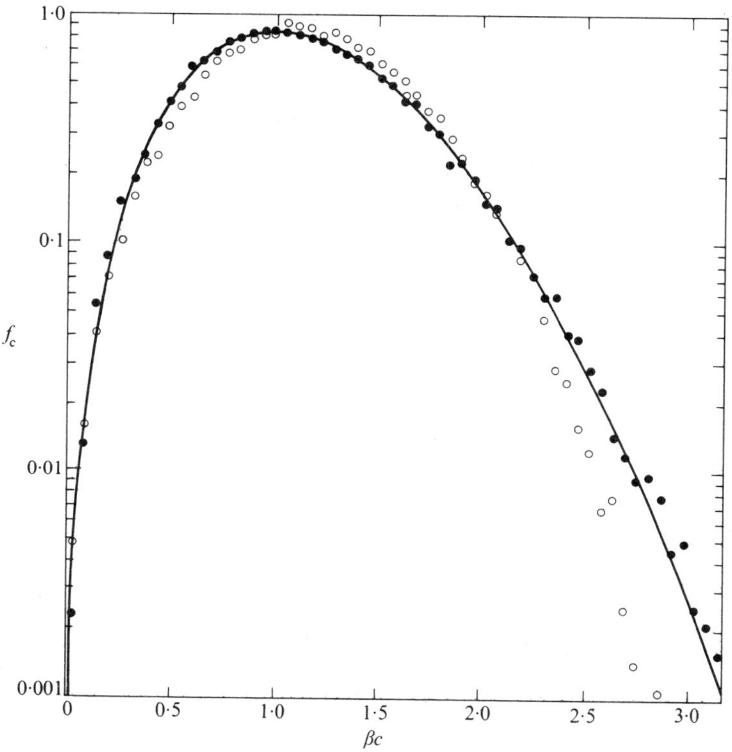

FIG. 11.3. Molecular speed distribution function in an equilibrium, homogeneous, stationary gas. —— Maxwellian distribution (eqn (4.6)); ● Direct simulation result for rough-sphere molecules; ○ Direct simulation result for the energy sink model with $\zeta = 3, \eta = \infty$, and $t_f = 0{\cdot}26$.

model with three internal degrees of freedom matches the relaxation rate of rough-sphere molecules when the transfer factor is equal to 0·26. However, there was a suggestion that the rotational temperature does not quite reach the equilibrium value when t_f is as large as 0·26. The reason for this is provided by Fig. 11.3 which compares the equilibrium translational speed distributions for the rough-sphere and energy-sink model gases. The rough-sphere distribution fits the Maxwellian curve to within the expected scatter, but there is a significant defect in the high-speed tail of the energy-sink distribution. This occurs for molecular speeds in excess of twice the most probable speed and the question arises as to whether it would lead to significant error in applications of the energy-sink model. Both models were used to calculate shock-wave profiles by the direct-simulation Monte Carlo method for a shock Mach number of three. No significant differences were found in the results. The breakdown of translational and rotational equilibrium in a steady expansion would be expected to provide a more severe

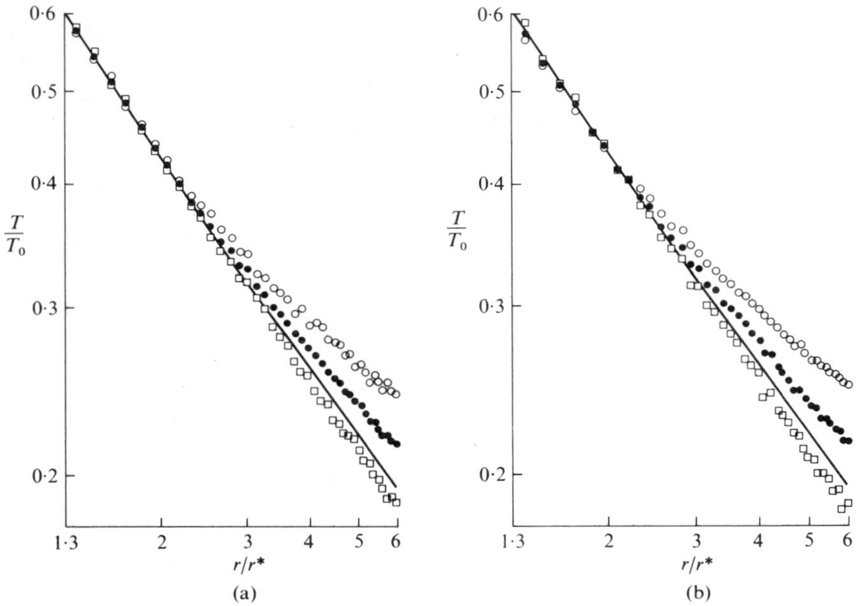

FIG. 11.4. Breakdown of equilibrium in a steady sphereical expansion with $(Kn)^* = \lambda_0/r^* =$ 0·002. —— Continuum result for T/T_0; ●, ○, and □ Direct simulation results for T_{0v}/T_0, T_{int}/T_0, and T_{tr}/T_0, respectively, for (a) Rough-sphere molecules, and (b) Energy sink model with $\zeta = 3$, $\eta = \infty$, and $t_f = 0.26$.

test than that posed by the shock wave calculation. Comparative calculations were therefore made for a spherical expansion with a throat radius equal to 500 stagnation free paths. Fig. 11.4 shows the ratios of the kinetic temperatures to the stagnation temperature T_0 as a function of the radius r normalized to the throat radius r^*. In addition to the overall temperature T_{ov} results are shown for the translational temperature T_{tr} and the internal temperature T_{int}. The radial and longitudinal temperatures are not shown, but the divergence of these coincided with that of the internal and translational temperatures. As in the shock wave calculation, the energy-sink model provides an effective duplication of the rough-sphere result.

Simulation calculations employing the energy sink model have also been compared with experiment. As an example, Fig. 11.5 presents the density and temperature profiles for a shock wave of Mach number 1·71. The experimental profiles were measured in nitrogen by Robben and Talbot (1966), while the direct simulation computations were for inverse ninth power molecules with two internal degrees of freedom. The transfer factor of 0·046 corresponds to a rotational relaxation time of five mean collision times. Since this factor is much smaller than the 0·26 required for simulation

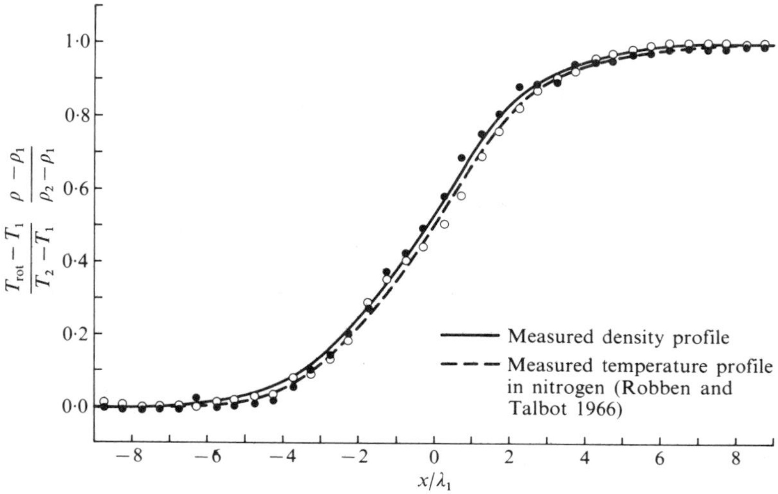

FIG. 11.5. Comparison of experimental and numerical results for the density and temperature profiles in a shock wave of Mach number 1·71 in a diatomic gas. Direct simulation results for the density and temperature profiles using the energy sink model with $\zeta = 2$, $\eta = 9$, and $t_f = 0{\cdot}046$.

of the rough sphere model, the distortion of the equilibrium distribution function is negligible. The results indicate that the energy sink model can provide an adequate representation of non-equilibrium flows with internal degrees of freedom. Similar comparison have been made by Larsen and Borgnakke (1973, 1974) using their statistical collision model. Since their model leads unequivocally to the Maxwellian distribution in equilibrium under all circumstances, it may be preferred for many applications. These would certainly include simulations which involve chemical reactions that occur only as a result of the more energetic collisions. On the other hand, the simpler energy sink model would be preferred when direct simulation is used in the continuum regime. This model was, in fact, used for the jet-plume impingement study that led to Fig. 9.5.

12

CHEMICALLY REACTING GAS FLOWS

12.1. Collision theory for reaction rates

MOST gas-phase chemical reactions may be treated as collision processes. We have seen that the less complicated of the classical molecular models provide an adequate representation of monatomic gas flows, and that simple modifications of these lead to acceptable results for gases with internal degrees of freedom. We will now investigate the extent to which this approach can be extended to deal with flows that involve chemical reactions. A logical starting point is provided by the classical collision theory that has long been established as the simplest theory for reaction rates.

The dilute-gas model deals exclusively with binary collisions and is most readily extended to *bimolecular* reactions. A typical bimolecular reaction may be written schematically as

$$\mathrm{A}+\mathrm{B} \rightleftarrows \mathrm{C}+\mathrm{D}, \tag{12.1}$$

where A, B, C, and D represent separate molecular species. As in the earlier chapters, the word 'molecule' is used as a general term and includes atoms, ions, electrons, and photons, as well as molecules. As long as the reaction takes place in a single step with no species other than the reactants present, as is implied by eqn (12.1), it has been found that the rate of change of concentration of species A may be written

$$-\frac{\mathrm{d}n_{\mathrm{A}}}{\mathrm{d}t} = k_{\mathrm{f}}(T)n_{\mathrm{A}}n_{\mathrm{B}} - k_{\mathrm{r}}(T)n_{\mathrm{C}}n_{\mathrm{D}}. \tag{12.2}$$

The number density has been used as the measure of concentration instead of the more usual moles per unit volume, denoted by $[N_{\mathrm{A}}]$. The two are related by $n_{\mathrm{A}} = \mathcal{N}[N_{\mathrm{A}}]$, where $\mathcal{N}$ is Avogardro's number. The rate coefficients or 'constants' $k_{\mathrm{f}}(T)$ and $k_{\mathrm{r}}(T)$ are for the forward and reverse reactions, respectively. These are functions of temperature and are independent of the number densities and time. A further result that originally had an empirical basis is that the rate coefficients are of the form

$$k(T) = \Lambda T^{\eta} \exp(-E_{\mathrm{a}}/kT), \tag{12.3}$$

where Λ and η are constants, and E_a is called the activation energy of the reaction. For the special case of $\eta = 0$, this equation is called the Arrhenius equation. The formulation is essentially macroscopic, and the use of the thermodynamic temperature implies that the reaction is a quasi-equilibrium process. This requires that the time scale of the reaction process should be large compared with the mean collision time, thus enabling the distributions of the translational and rotational molecular velocities to remain essentially Maxwellian.

Although the total cross-section σ_T was originally defined for elastic collisions, it generally does not require any modification for application to collisions that involve energy interchange between translational and fully excited internal modes. However, for chemical reactions or for the excitation of discrete vibrational energy levels, a separate *reaction cross-section* σ_R is introduced. This is a total cross-section, but a more complete knowledge of a reaction would enable a differential cross-section to be specified as a function of the impact parameters. The ratio σ_R/σ_T of the reaction cross-section to the elastic cross-section may be regarded as the probability that an elastic collision will result in a chemical reaction. It may then be identified with the *steric factor*. However, the use of a steric factor implies that the reaction cross-section must be smaller than the elastic cross-section and, while this is generally the case, it is desirable to leave open the possibility that $\sigma_R > \sigma_T$. If the bimolecular reaction between A and B requires a finite heat input, it is reasonable to assume that the reaction cannot occur unless the energy of the colliding molecules exceeds some threshold value E_a in the centre of mass frame of reference. If we consider only the relative translational energy along the line of centres and assume quasi-equilibrium conditions, eqn (4.51) states that the fraction of collisions in a hard sphere gas with the required energy is simply $\exp(-E_a/kT)$. Then, if the reaction cross-section is zero for collisions having less than the required energy, and equal to a constant value σ_R for collisions with the required energy, the forward reaction rate for molecule A is obtained by multiplying the elastic collision rate from eqn (4.39) by the above fraction and by the ratio σ_R/σ_T. This gives

$$\frac{dn_A}{dt} = -\sigma_R n_A n_B \left(\frac{8kT}{\pi m_r}\right)^{\frac{1}{2}} \exp\left(\frac{E_a}{kT}\right). \tag{12.4}$$

A comparison of this equation with eqns (12.2) and (12.3) shows that the collision theory result is of the expected form.

A more realistic result is obtained (Ross, Light, and Shuler 1969) if σ_R is a function of the amount by which the collision energy exceeds the threshold value E_a. The most convenient assumption is that it is proportional to the excess energy raised to the *j*th power. If we again consider the energy along the line of centres, the formal description of the reaction cross-section is

$$\sigma_R = 0 \qquad \text{for } \tfrac{1}{2}m_r c_r^2 < E_a,$$

and (12.5)

$$\sigma_R = \sigma_0 \left(1 - \frac{2E_a}{m_r c_r^2}\right)\left(\frac{m_r c_r^2}{2E_a} - 1\right)^j, \qquad \text{for } \tfrac{1}{2}m_r c_r^2 > E_a,$$

where σ_0 is a constant. Eqns (4.39) and (4.49) then lead to

$$\frac{dn_A}{dt} = -\sigma_0 n_A n_B \left(\frac{8kT}{\pi m_r}\right)^{\frac{1}{2}} \frac{2m_r^2}{(2kT)^2} \int_{(2E_a/m_r)^{1/2}}^{\infty} c_r^3 \left(1 - \frac{2E_a}{m_r c_r^2}\right) \times$$
$$\times \left(\frac{m_r c_r^2}{2E_a} - 1\right)^j \exp\left(-m_r c_r^2/(2kT)\right) dc_r.$$

This may be evaluated through eqns (B2) and (B4) to give the remarkably simple result

$$dn_A/dt = -\sigma_0 n_A n_B (8kT/\pi m_r)^{\frac{1}{2}} \Gamma(j+2)\,(kT/E_a)^j \exp(-E_a/kT). \quad (12.6)$$

Therefore, for this model, the constants in eqn (12.3) for the rate coefficient are

$$\Lambda = \Gamma(j+2)\sigma_0 (8k/\pi m_r)^{\frac{1}{2}} (k/E_a)^j,$$

and (12.7)

$$\eta = j + \tfrac{1}{2}.$$

The agreement between the accepted formulation for the reaction-rate coefficients and the general results from the collision theory indicates that there is some physical justification for the model. However, a direct deduction of physical quantities from relations such as eqn (12.7) may be misleading. While the simple form of eqn (12.6) makes the reaction cross-section of eqn (12.5) a very attractive model, the physics of the reaction cannot be expected to be governed by the mathematical peculiarities of the incomplete gamma function. Apart from the classical approximation to the actual quantum processes and the averaging that is implied by the use of the total cross-section, the major deficiency of the simple theory is related to the role played by the internal degrees of freedom in the reaction. The usual procedure for dealing with this problem is to introduce the concept of effective 'square terms' (see, for example, Vincenti and Kruger (1965), Chapter VII, § 5). The translational mode contributes two square terms and the total number may be denoted by $2s$. When used in conjunction with the simple form of the reaction cross-section that led to eqn (12.4), and assuming that $E_a \gg (s-1)kT$, the 'square term theory' leads to an equation of the form of the generalized Arrhenius equation, but with $\eta = \tfrac{3}{2} - s$. This relation has been compared with experimentally determined values of η to assess the effectiveness of internal degrees of freedom. This is unlikely to lead to meaningful conclusions, since eqn (4.47) indicates that η should be reduced by approximately $\frac{1}{4}$ for realistically soft molecules and there will almost certainly be some effect due to the energy dependence of the reactive cross-section, as indicated by eqn (12.7). However, the major objection to the square term concept in the present context is that it is based on statistical mechanics rather than the mechanics of the individual collisions. This means that it could not be readily incorporated into a direct simulation approach.

It may be argued that contributions from internal modes could enable the reaction to proceed even though the relative translational energy was less than the required threshold value E_a. A direct application of this idea would lead to a lower effective value of the activation energy in the exponential term of the Arrhenius equation. In support of this, it has been found (e.g., Giedt, Cohen, and Jacobs 1969) that the activation energy for diatomic molecules may be one or two multiples of kT less than the bond dissociation energy. Therefore, for dissociation reactions, the energy of dissociation in the collision theory may be less than the bond dissociation energy. Alternatively, it may be argued that the contribution of internal degrees of freedom would be analogous to allowing some translational energy, other than that along the line of centres, to contribute to the available energy. For example, a comparison of eqns (4.49) and (4.51) shows that the complete removal of the line of centres requirement would lead to the inclusion of the additional factor $1+(E_a/kT)$ on the right hand side of eqns (12.4) and (12.6). This is equivalent to the contribution of two square terms by the internal degrees of freedom.

The quasi-equilibrium collision theory may be extended to *ternary* reactions which involve triple collisions. The recombination reaction falls into this class, since a third body must be involved in the collision in order to carry off part of the energy released in the recombination. The simplest approach is to assign a 'lifetime' to a binary collision and to compute triple collisions as binary collisions between the pair of molecules in the initial collision and a third molecule. While this general approach has been adopted by a number of workers (Smith 1969), the following analysis is more precisely related to kinetic theory and may be readily implemented in a direct simulation approach.

Consider triple or three-body collisions between a type A molecule and a collision pair consisting of a type B and a type C molecule. The number density of BC collision pairs is equal to the product of the number of collisions between type B and C molecules per unit volume per unit time with the lifetime of a collision. The collision lifetime is assumed to have the form

$$a_{tBC}\sigma_{BC}^{\frac{1}{2}}/c_r, \tag{12.8}$$

where a_{tBC} is a constant, σ_{BC} is the total cross-section, and c_r is the relative speed. Now consider collisions of a type B test molecule with the class of type C molecules such that the relative speed is c_r. Eqn (1.6) shows that the collision rate of the test molecule is $\mathrm{d}n_C\sigma_{BC}c_r$, where $\mathrm{d}n_C$ is the number density of the class of type C molecules. Then, if $\mathrm{d}n_B$ is the number density of the type B molecules in the same velocity class as the test molecule, the contribution of this class of collisions to the number density n_{BC} of BC collision pairs is

$$\begin{aligned}\mathrm{d}n_{BC} &= \mathrm{d}n_B\mathrm{d}n_C\sigma_{BC}c_r a_{tBC}\sigma_{BC}^{\frac{1}{2}}/c_r \\ &= a_{tBC}\sigma_{BC}^{\frac{3}{2}}\,\mathrm{d}n_B\,\mathrm{d}n_C.\end{aligned}$$

Since this result is independent of the relative speed, the required number density is simply

$$n_{BC} = a_{tBC}\sigma_{BC}^{\frac{3}{2}}n_B n_C/\varepsilon. \tag{12.9}$$

The symmetry factor ε is included to allow for the double counting of collisions when B = C. The formal definition is

$$\varepsilon = 2 \qquad \text{for B} = \text{C},$$

and

$$\varepsilon = 1 \qquad \text{for B} \neq \text{C}.$$

The pair of molecules moves with the centre of mass velocity $\boldsymbol{c}_{mBC}$. The equilibrium distribution function for $\boldsymbol{c}_{mBC}$ is, from eqn (4.33),

$$f c_{mBC} = \frac{4}{\pi^{\frac{1}{2}}}\left(\frac{m_B+m_C}{2kT}\right)^{\frac{3}{2}} c_{mBC}^2 \exp\left\{-\frac{(m_B+m_C)c_{mBC}^2}{2kT}\right\}. \tag{12.10}$$

There can be no favoured direction and a comparison of eqns (12.10) and (4.6) shows that the distribution function for $\boldsymbol{c}_{mBC}$ is the Maxwellian corresponding to a molecule of mass m_B+m_C. The equilibrium collision rate for type A molecules with the BC pairs then follows from eqn (4.39) as

$$\frac{2}{\pi^{\frac{1}{2}}}\sigma_{A(BC)}n_A n_{BC}\left\{\frac{2kT(m_A+m_B+m_C)}{m_A(m_B+m_C)}\right\}^{\frac{1}{2}}$$

or, substituting from eqn (12.9),

$$\frac{a_{tBC}}{\varepsilon}\sigma_{A(BC)}\sigma_{BC}^{\frac{3}{2}}n_A n_B n_C\left\{\frac{8kT(m_A+m_B+m_C)}{\pi m_A(m_B+m_C)}\right\}^{\frac{1}{2}}. \tag{12.11}$$

Ternary reaction rates may be related to this triple collision rate, just as bimolecular collision rates were related to the binary collision rate of eqn (4.39). The theory could be extended to determine the fraction of collisions with relative translational energies above a certain limit. However, the important ternary reactions are exothermic, rather than endothermic, and the available evidence suggests that the activation energy should then be set to zero and the reactive cross-section set equal to the elastic cross-section.

12.2. The dissociation–recombination reaction

The dissociation–recombination of a diatomic gas is one of the most important, and certainly the most widely studied reaction in gas dynamics. For a homogeneous pure gas, it may be written

$$\mathrm{M} + \left\{\begin{matrix}\mathrm{M}\\ \mathrm{A}\end{matrix}\right\} \rightleftarrows \mathrm{A} + \mathrm{A} + \left\{\begin{matrix}\mathrm{M}\\ \mathrm{A}\end{matrix}\right\}, \tag{12.12}$$

where M denotes the molecule and A denotes the atom. The forward or dissociation reaction is bimolecular, while the backward or recombination reaction is termolecular. The collision partner of the molecule in the dissociation reaction may be either a molecule or an atom.

The rate of loss of molecules due to dissociation reactions with an energy of dissociation E_d may be written directly from eqn (12.6) as

$$-\frac{\mathrm{d}n_\mathrm{M}}{\mathrm{d}t} = \tfrac{1}{2}n_\mathrm{M}(\sigma_{0,\mathrm{MM}}n_\mathrm{M} + 6^{\frac{1}{2}}\sigma_{0,\mathrm{MA}}n_\mathrm{A})\left(\frac{8kT}{\pi m_\mathrm{A}}\right)^{\frac{1}{2}}\Gamma(j+2)\left(\frac{kT}{E_\mathrm{d}}\right)^{j}\exp\left(-\frac{E_\mathrm{d}}{kT}\right). \tag{12.13}$$

The numerical factors have been affected by the application of the symmetry factor of $\frac{1}{2}$ to the molecule–molecule collisions and the the relationship of the reduced masses to the mass of an atom.

For recombination reactions, it is assumed that there is no difference in recombination effectiveness between the collision of an atom with an atom–molecule pair and the collision of a molecule with an atom–atom pair. The collision of atoms with atom–atom pairs must also be considered. With the further assumption that a recombination results from every ternary collision that involves at least two atoms, the rate of formation of molecules follows from eqn (12.11) as

$$\frac{\mathrm{d}n_\mathrm{M}}{\mathrm{d}t} = n_\mathrm{A}^2[n_\mathrm{M}\{\tfrac{1}{2}a_{\mathrm{t,AA}}\sigma_{\mathrm{M(AA)}}\sigma_{\mathrm{AA}}^{\frac{3}{2}} + (2/3^{\frac{1}{2}})a_{\mathrm{t,AM}}\sigma_{\mathrm{A(AM)}}\sigma_{\mathrm{AM}}^{\frac{3}{2}}\} +$$

$$+ (3/8)^{\frac{1}{2}}n_\mathrm{A}a_{\mathrm{t,AA}}\sigma_{\mathrm{A(AA)}}\sigma_{\mathrm{AA}}^{\frac{3}{2}}]\left(\frac{8kT}{\pi m_\mathrm{A}}\right)^{\frac{1}{2}}. \tag{12.14}$$

In equilibrium, the rate of formation of atoms must be exactly balanced by the corresponding rate of loss and the two expressions may be equated. The resulting expression is greatly simplified for particular relationships between the cross-sections and collision lifetimes. A number of plausible assumptions may be made, one set being

$$\tfrac{1}{2}\sigma_{0,\mathrm{MM}} = 6^{\frac{1}{2}}\sigma_{0,\mathrm{MA}} = \sigma_0,$$
$$a_{t,\mathrm{AM}}\sigma_{\mathrm{AM}}^{\frac{3}{2}} = (3^{\frac{1}{2}}/4)a_{t,\mathrm{AA}}\sigma_{\mathrm{AA}}^{\frac{3}{2}}, \tag{12.15}$$

and

$$\sigma_{\mathrm{A(AM)}} = \sigma_{\mathrm{M(AA)}} = (\tfrac{3}{2})^{\frac{1}{2}}\sigma_{\mathrm{A(AA)}} = 2\sigma_{\mathrm{AA}}.$$

The equilibrium condition then becomes

$$n_{\mathrm{A}}^2 a_{t,\mathrm{AA}}\sigma_{\mathrm{AA}}^{\frac{5}{2}} = \tfrac{1}{2}n_{\mathrm{M}}\sigma_0\Gamma(j+2)\left(\frac{kT}{E_{\mathrm{d}}}\right)^j \exp\left(-\frac{E_{\mathrm{d}}}{kT}\right). \tag{12.16}$$

The *degree of dissociation* α may now be defined as the mass fraction of dissociated gas. Therefore,

$$\alpha = \frac{n_{\mathrm{A}}m_{\mathrm{A}}}{\rho} = \frac{n_{\mathrm{A}}}{2n_{\mathrm{M}}+n_{\mathrm{A}}} \tag{12.17}$$

and the equilibrium condition is

$$\frac{\alpha^2}{1-\alpha} = \frac{m_{\mathrm{A}}}{\rho}\frac{\sigma_0\Gamma(j+2)}{4a_{t,\mathrm{AA}}\sigma_{\mathrm{AA}}^{\frac{5}{2}}}\left(\frac{kT}{E_{\mathrm{d}}}\right)^j \exp\left(-\frac{E_{\mathrm{d}}}{kT}\right). \tag{12.18}$$

The form of this equation corresponds exactly with the standard result from statistical mechanics. Vincenti and Kruger (1965) give the following form of the law of mass action for this reaction:

$$\frac{\alpha^2}{1-\alpha} = \frac{m_{\mathrm{A}}}{\rho}\frac{(Q^a)^2}{2VQ^{aa}}\exp\left(-\frac{E_{\mathrm{d}}}{kT}\right). \tag{12.19}$$

The partition functions Q^a and Q^{aa} each contain the volume V as a multiplying factor, so that this quantity disappears and the right hand side is essentially a function of temperature divided by the density. The partition functions contain Planck's constant which is, for course, inaccessible to the classical theory that leads to eqn (12.18). Note that the T^j term in eqn (12.18) enters through the reactive cross-section of eqn (12.5) and that alternative forms of this cross-section, or modifications to the theory to allow for the effects of internal degrees of freedom, would alter the form of this temperature dependence.

12.3. Simulation of reacting gas flows

The direct simulation Monte Carlo method permits the computation of nonlinear and nonequilibrium flow fields to the extent that physical information is available on the molecular collision processes. As far as chemical reactions are concerned, the ideal situation would be to have complete information on the reactive cross-sections as functions of the geometric impact parameters, the relative translational energy, and the molecular orientations and internal states. The information that is available generally presents averaged cross-sections as a function of the relative energy. However, the cross-sections are frequently dependent on the vibrational level of the molecules and additional cross-sections may be established for collisional transfers from one level to another. As long as the number of populated vibrational levels is reasonably small, the molecules in each level may be treated as separate species of a gas mixture, as described in § 11.1. The capability of handling a mixture containing a large number of components is a major advantage of the simulation approach. Apart from this aspect, the paucity of the information on reactive cross-sections necessitates the adoption of simulation procedures that are closely related to the collision theory analysis of the preceding sections. While the collision theory is restricted to quasi-equilibrium reactions in a spatially homogeneous gas, the simulation approach enables consideration to be given to the effects produced by large gradients in the flow at the macroscopic level and by large departures from the equilibrium velocity distribution at the microscopic level.

A Monte Carlo simulation method has been used by Koura (1973) to study the extent of non-equilibrium effects in a typical chemical reaction in a spatially homogeneous gas. While a number of cases were considered, the study points up one of the major difficulties associated with the application of any numerical method to problems that progressively become more complex. This is that the parameter space becomes so large that it is difficult to study a sufficient number of cases to establish the overall character of the solution. The interpretation of the results from a limited number of particular cases is considerably facilitated if the simulation is based on procedures that permit an analytical solution in the homogeneous quasi-equilibrium gas case. This is because the collision cross-section and lifetimes that are used in the simulation of collisions may then be related to reaction rates and equilibrium conditions, respectively. This is particularly important in the case of the collision lifetimes, since they cannot be directly related to real physical quantities. The theory enables the reactive cross-sections to be inferred from, or at least made consistent with, experimental data on reaction rates. However, it should be kept in mind that the chemical reaction may distort the distribution function to such an extent that the quasi-equilibrium reaction rate would not provide a reliable result. A future application of the simulation method will be to provide information on this point.

We will now describe how the theoretical models that were used in the dissociation–recombination theory of §§ 12.1 and 12.2 may be implemented in a simulation program. Any simulation that involves chemical reactions must necessarily deal with a gas mixture, so that the extensions are to the procedures of Chapter 10. While it is necessary to modify the overall collision procedures, the basic collision mechanics may be based on either the hard-sphere or the more general inverse power law model. The internal degrees of freedom will generaly be dealt with through a phenomenological model, although molecules in different vibrational states may be regarded as distinct species. Each time a binary collision is computed, the relevant reactive cross-sections are calculated from equations such as (12.5) and the ratios of these to the elastic cross-sections give the probabilities of the various reactions. The standard acceptance–rejection method is then used to decide whether or not a reaction occurs. The procedures which conserve momentum while allowing for changes in the translational energy have already been described in the context of the energy sink model. The only additional action that is required is to change the integer which indicates the chemical nature of the particle or particles that undergo chemical change and, in the case of a dissociation, to add the additional particle. More extensive modifications to the basic procedures are required for the simulation of the triple collisions. Just as separate time counters are used for each cell and for each class of collision in a gas mixture, separate counters may be specified for each of the triple collisions that involve at least two atoms. The computation of the triple collisions is best dealt with quite separately from that of the binary collisions. When the triple-collision rate is one or more orders of magnitude less than the binary collision rate, a separate and proportionately larger time interval Δt_m may be introduced for the triple collisions. The triple collision is regarded as a binary collision between molecule A and a binary collision pair comprising molecules B and C. The lifetime of the collision pair is given by eqn (12.8) and, since this is inversely proportional to relative speed between B and C, the pair may be chosen at random and the three molecules are accepted or rejected according to the relative speed c_r between molecule A and the pair BC. In calculating this relative speed, the appropriate velocity of the pair is, of course, the centre of mass velocity. The time increment for the collision is given by eqn (10.3) as

$$\Delta t_c = \frac{1}{N_A}\frac{1}{\sigma_{A(BC)} n_{BC} c_r} + \frac{1}{N_{BC}}\frac{1}{\sigma_{A(BC)} n_A c_r}. \tag{12.20}$$

The number density n_{BC} is given by eqn (12.9) and effective value of N_{BC} follows from the relation

$$\frac{N_{BC} W_{f_{BC}}}{N_A W_{f_A}} = \frac{n_{BC}}{n_A}, \tag{12.21}$$

where W_f denotes the weighting factor. Note that the two components of eqn (12.20) are equal in the absence of weighting factors. Otherwise, the components are subject to probability factors based on the weighting factors as was described in § 10.2 in connection with eqn (10.3). The various collision cross-sections and lifetimes are related through eqn (12.15). The centre of mass velocity of the three pre-collision particles is calculated and this is retained as the centre of mass velocity of the two post-collision particles. The recombination energy is added to the pre-collision translational energy and the total energy may be divided between the two post-collision particles on the assumption that the direction of the post-collision relative velocity may be chosen at random. This effectively assumes hard-sphere collision mechanics for the triple collision.

The dissociation–recombination reaction has been simulated for the special case of a homogeneous gas in which all the energy is initially in the translational and rotational modes. The degree of dissociation was sampled as a function of time. The resulting curves were of the typical relaxation form, with the initial slopes being consistent with the reaction rate prediction of eqn (12.6). The asymptotic level of the degree of dissociation was consistent with the equilibrium prediction of eqn (12.19).

When the gas temperature is such that the ratio kT/E_a is small, reactions with cross-sections similar to that of eqn (12.5) occur only as a result of collisions between molecules from the extreme high-speed tail of the speed distribution. Since these constitute a very small fraction of the total molecular sample, severe fluctuation problems may result from the small effective sample size. We have seen that weighting factors may be used to reduce sample inequalities in physical space. This principle may be extended to velocity space, with the molecules being assigned weighting factors that are inversely proportional to some power of the molecular speed. As before, the number of real molecules represented by each simulated molecule is proportional to its weighting factor. Molecules are duplicated or removed from the calculation as they move from one location to another in velocity space. Also the relative magnitudes of the weighting factors again determine whether or not the velocity components of a particular molecule are modified in a collision. The principle has been successfully tested in a program for translational relaxation in a homogeneous gas. The weighting factors were chosen such that equilibrium number of molecules fell as $\exp(-c)$, rather than $\exp(-c^2)$.

With the establishment of routines for handling chemical reactions, the direct simulation Monte Carlo method is capable of providing solutions for virtually all neutral gas flows that involve large perturbations and have minimum Knudsen numbers that are not less than 10^{-2} or 10^{-3}. The extent to which the capability is utilized will depend, to a large degree, on the relative status that is accorded to simulation studies, to experiments, and to theoretical analyses. We conclude with several observations on this point:

(*i*) Simulation results always reflect the deficiencies of the physical model on which they are based. No amount of simulation will, by itself, provide basic physical information. At the same time, the simulation method can provide information on flows that are not amenable to analysis and for which experiment is either impracticable or impossible.

(*ii*) The reason for much of the uncertainty associated with the currently available physical information is that many of the basic quantities are inferred from indirect measurements through questionable assumptions that form the basis of approximate and inadequate theories. If simulation studies were carried out in conjunction with the experiments, many of the assumptions and approximations would no longer be required, thus avoiding much of the error.

(*iii*) The above observations show that simulation and experiment should generally be regarded as complementary rather than competitive. The simulation usually requires so little effort, in comparison with the experiment, that consideration should be given to its use whenever the physical quantities of interest cannot be measured directly.

(*iv*) Nonlinear problems with complex boundary conditions are inaccessible to exact theoretical analysis. Simulation of these flows can provide a test of the assumptions on which approximate solutions are based, or can suggest alternative approximations that lead to new theories. To this extent, simulation is complementary to analysis as well as to experiment. However, when solutions are required for the *ad hoc* problems that are encountered in practice, simulation will frequently offer the only viable approach.

APPENDIX A

Representative gas properties

TABLE A1

Properties under standard conditions (101 325 Pa and 0 °C)

Gas	Symbol	Molecular weight ($\mathscr{M}$)	Ratio of specific heats (γ)	Nominal diameter† ($d \times 10^8$ cm)
Hydrogen	H_2	2·016	1·41	2·75
Helium	He	4·003	1·67	2·19
Methane	CH_4	16·04	1·31	4·19
Ammonia	NH_3	17·03	1·32	4·48
Neon	Ne	20·18	1·67	2·60
Carbon monoxide	CO	28·01	1·40	3·81
Nitrogen	N_2	28·01	1·40	3·78
Oxygen	O_2	32·00	1·40	3·64
Argon	Ar	39·94	1·67	3·66
Carbon dioxide	CO_2	44·01	1·30	4·64
Chlorine	Cl_2	70·91	1·33	5·53
Krypton	Kr	83·80	1·67	4·20
Freon 12	CCl_2F_2	120·9	1·13	6·41
Xenon	Xe	131·3	1·67	4·94

† Based on the measured coefficient of viscosity and the Chapman–Enskog result of eqn (4.52) for hard sphere molecules.

Values for the physical constants are given in the List of Symbols. These may be combined with the gas properties listed in Table A1 to give other properties, as follows: Molecular mass

$$m = \mathscr{M}/\mathscr{N} = 1{\cdot}660 \times 10^{-24}\,\mathscr{M}\ \mathrm{g} = 1{\cdot}660 \times 10^{-27}\,\mathscr{M}\ \mathrm{kg} \qquad \text{(A1)}$$

Density under standard conditions (101 325 Pa and 0 °C)

$$\rho_0 = n_0 m = 4{\cdot}462 \times 10^{-5}\,\mathscr{M}\ \mathrm{g\,cm^{-3}} = 0{\cdot}04462\,\mathscr{M}\ \mathrm{kg\,m^{-3}} \qquad \text{(A2)}$$

Gas constant

$$R = \mathscr{R}/\mathscr{M} = 8{\cdot}3143/\mathscr{M}\ \mathrm{J\,mol^{-1}\,K^{-1}} = 8314{\cdot}3/\mathscr{M}\ \mathrm{m^2\,s^{-2}\,J^{-1}} \qquad \text{(A3)}$$

Speed of sound (at a temperature T K)

$$a = (\gamma R T)^{\frac{1}{2}} = 91{\cdot}18\,(\gamma T/\mathscr{M})^{\frac{1}{2}}\ \mathrm{m\,s^{-1}} \qquad \text{(A4)}$$

Most probable molecular speed in an equilibrium gas (T in K)

$$c_m = (2RT)^{\frac{1}{2}} = 128{\cdot}95\,(T/\mathscr{M})^{\frac{1}{2}} \text{ m s}^{-1} \tag{A5}$$

Mean free path in an equilibrium gas at density n (d in cm)

$$\lambda_0 = (2^{\frac{1}{2}}\pi d^2 n)^{-1} = 8{\cdot}3766 \times 10^{-21}\,(n_0/n)d^{-2} \text{ cm} \tag{A6}$$

Effective number of internal degrees of freedom

$$\zeta = (5-3\gamma)/(\gamma-1) \tag{A7}$$

APPENDIX B
Definite integrals

The taking of moments of the equilibrium distribution function leads to definite integrals of the form

$$\int_a^\infty v^n \exp(-\beta^2 v^2)\, dv. \tag{B1}$$

The incomplete gamma function is defined by

$$\Gamma(j, \alpha) = \int_\alpha^\infty x^{j-1} e^{-x}\, dx \tag{B2}$$

and it is readily seen that

$$\Gamma\left(\frac{n+1}{2}, \beta^2 a^2\right) = 2\beta^{n+1} \int_{|a|}^\infty v^n \exp(-\beta^2 v^2)\, dv. \tag{B3}$$

The integral of eqn (B1) may therefore be reduced through the following standard theorem

$$\Gamma(j, \alpha) = (j-1)\Gamma(j-1, \alpha) + \alpha^{j-1} e^{-\alpha}. \tag{B4}$$

Note that

$$\Gamma(0, \alpha) = E_1(\alpha),$$

where $E_1(\alpha)$ is the exponential integral and

$$\Gamma(\tfrac{1}{2}, \alpha) = \pi^{\frac{1}{2}} \operatorname{erfc}(\alpha^{\frac{1}{2}}),$$

where erfc is the complementary error function.

General results may be written down for the special case in which n is an integer. It is desirable to distinguish between odd and even values of n and between positive and negative values of a. We then have

$$\int_{\pm a}^\infty v^{2n} \exp(-\beta^2 v^2)\, dv = \frac{(2n-1)(2n-3)\ldots 1}{2^{n+1}\beta^{2n+1}} \pi^{\frac{1}{2}}\{1 \mp \operatorname{erf}(\beta a)\} \pm$$
$$\pm \frac{(\beta a)\exp(-\beta^2 a^2)}{\beta^{2n+1}} \sum_{m=1}^{n} \left(\frac{1}{2^m} \frac{(2n-1)(2n-3)\ldots 1}{\{2(n+1-m)-1\}\{2(n+1-m)-3\}\ldots 1} (\beta a)^{2(n-m)}\right) \tag{B5}$$

and, noting that $0! = 1$,

$$\int_{\pm a}^\infty v^{(2n+1)} \exp(-\beta^2 v^2)\, dv = \frac{\exp(-\beta^2 a^2)}{2\beta^{2n+2}} \sum_{m=1}^{n+1} \left\{\frac{n!}{(n+1-m)!} (\beta a)^{2(n+1-m)}\right\}. \tag{B6}$$

For $n = 0$, 1, and 2 these equations give:

$$\int_{\pm a}^\infty \exp(-\beta^2 v^2)\, dv = \frac{\pi^{\frac{1}{2}}}{2\beta}\{1 \mp \operatorname{erf}(\beta a)\} \tag{B7}$$

$$\int_{\pm a}^{\infty} v \exp(-\beta^2 v^2)\,\mathrm{d}v = \frac{\exp(-\beta^2 a^2)}{2\beta^2} \tag{B8}$$

$$\int_{\pm a}^{\infty} v^2 \exp(-\beta^2 v^2)\,\mathrm{d}v = \frac{\pi^{\frac{1}{2}}}{4\beta^3}\{1 \mp \operatorname{erf}(\beta a)\} \pm \frac{\beta a \exp(-\beta^2 a^2)}{2\beta^3} \tag{B9}$$

$$\int_{\pm a}^{\infty} v^3 \exp(-\beta^2 v^2)\,\mathrm{d}v = \frac{\exp(-\beta^2 a^2)}{2\beta^4}(1+\beta^2 a^2) \tag{B10}$$

$$\int_{\pm a}^{\infty} v^4 \exp(-\beta^2 v^2)\,\mathrm{d}v = \frac{3\pi^{\frac{1}{2}}}{8\beta^5}\{1 \mp \operatorname{erf}(\beta a)\} \pm \frac{\beta a \exp(-\beta^2 a^2)}{2\beta^5}\left(\frac{3}{2}+\beta^2 a^2\right) \tag{B11}$$

$$\int_{\pm a}^{\infty} v^5 \exp(-\beta^2 v^2)\,\mathrm{d}v = \frac{\exp(-\beta^2 a^2)}{2\beta^6}(2+2\beta^2 a^2+\beta^4 a^4) \tag{B12}$$

For the special case in which a is $-\infty$, eqns (B5) and (B6) show that

$$\int_{-\infty}^{\infty} v^{2n} \exp(-\beta^2 v^2)\,\mathrm{d}v = 2\int_{0}^{\infty} v^{2n} \exp(-\beta^2 v^2)\,\mathrm{d}v, \tag{B13}$$

and

$$\int_{-\infty}^{\infty} v^{(2n+1)} \exp(-\beta^2 v^2)\,\mathrm{d}v = 0. \tag{B14}$$

For $a = 0$, the results are related to the gamma function and eqns (B5) and (B6) become

$$\int_{0}^{\infty} v^{2n} \exp(-\beta^2 v^2)\,\mathrm{d}v = \frac{(2n-1)(2n-3)\ldots 1}{2^{n+1}\beta^{2n+1}}\pi^{\frac{1}{2}}, \tag{B15}$$

and

$$\int_{0}^{\infty} v^{2n+1} \exp(-\beta^2 v^2)\,\mathrm{d}v = \frac{n!}{2\beta^{2n+2}}. \tag{B16}$$

The results for $n = 0$ to 3 may be listed for quick reference as follows:

$$\int_{0}^{\infty} \exp(-\beta^2 v^2)\,\mathrm{d}v = \frac{\pi^{\frac{1}{2}}}{2\beta} \tag{B17}$$

$$\int_{0}^{\infty} v \exp(-\beta^2 v^2)\,\mathrm{d}v = \frac{1}{2\beta^2} \tag{B18}$$

$$\int_{0}^{\infty} v^2 \exp(-\beta^2 v^2)\,\mathrm{d}v = \frac{\pi^{\frac{1}{2}}}{4\beta^3} \tag{B19}$$

$$\int_{0}^{\infty} v^3 \exp(-\beta^2 v^2)\,\mathrm{d}v = \frac{1}{2\beta^4} \tag{B20}$$

$$\int_{0}^{\infty} v^4 \exp(-\beta^2 v^2)\,\mathrm{d}v = \frac{3\pi^{\frac{1}{2}}}{8\beta^5} \tag{B21}$$

$$\int_{0}^{\infty} v^5 \exp(-\beta^2 v^2)\,\mathrm{d}v = \frac{1}{\beta^6} \tag{B22}$$

$$\int_{0}^{\infty} v^6 \exp(-\beta^2 v^2)\,\mathrm{d}v = \frac{15\pi^{\frac{1}{2}}}{16\beta^7} \tag{B23}$$

$$\int_{0}^{\infty} v^7 \exp(-\beta^2 v^2)\,\mathrm{d}v = \frac{3}{\beta^8} \tag{B24}$$

APPENDIX C

The error function

The error function of the argument a is defined by

$$\operatorname{erf}(a) = \frac{2}{\pi^{\frac{1}{2}}} \int_0^a \exp(-x^2)\,dx \tag{C1}$$

and the complementary error function by

$$\operatorname{erfc}(a) = 1 - \operatorname{erf}(a). \tag{C2}$$

TABLE C1

Representative values of the error function

a	erf (a)	a	erf (a)
0·	0·0	1·1	0·880 205
0·05	0·056 372	1·2	0·910 314
0·1	0·112 463	1·3	0·934 008
0·15	0·167 996	1·4	0·952 285
0·2	0·222 703	1·5	0·966 105
0·25	0·276 326	1·6	0·976 348
0·3	0·328 627	1·7	0·983 790
0·35	0·379 382	1·8	0·989 091
0·4	0·428 392	1·9	0·992 790
		2·0	0·995 322
0·45	0·475 482		
0·5	0·520 500	2·2	0·998 137
0·55	0·563 323	2·4	0·999 311
0·6	0·603 856	2·6	0·999 764
		2·8	0·999 925 0
0·65	0·642 029	3·0	0·999 977 9
0·7	0·677 801		
0·75	0·711 156	3·2	0·999 993 97
0·8	0·742 101	3·4	0·999 998 48
		3·6	0·999 999 64
0·85	0·770 668	3·8	0·999 999 92
0·9	0·796 908	4·0	0·999 999 98
0·95	0·820 891		
1·0	0·842 701	∞	1·0

Note that

$$\operatorname{erf}(-a) = -\operatorname{erf}(a),$$

$$\operatorname{erf}(0) = 0, \tag{C3}$$

and

$$\operatorname{erf}(\infty) = 1.$$

The following asymptotic series is useful when large arguments are encountered

$$\operatorname{erf}(a) = 1 - \frac{1}{\pi^{\frac{1}{2}}} \exp(-a^2) \left(\frac{1}{a} - \frac{1}{2a^3} + \frac{1 \cdot 3}{2^2 a^5} - \dots \right). \tag{C4}$$

The most useful series for the computation of the error function is

$$\operatorname{erf}(a) = \frac{2}{\pi^{\frac{1}{2}}} \exp(-a^2) \sum_{n=0}^{\infty} \frac{2^n}{1 \cdot 3 \dots (2n+1)} a^{2n+1}. \tag{C5}$$

Since the ratio of the nth to the $(n-1)$th term is simply $2a^2/(2n+1)$, eqn (C5) is readily programmed as a computer subroutine. Alternatively, the error function may be computed from the rational approximations listed by Abramowitz and Stegun (1965).

APPENDIX D

Sampling from a prescribed distribution

Probabilistic modelling of physical processes requires the generation of representative values of variables that are distributed in a prescribed manner. This is done through random numbers and is a key step in direct simulation Monte Carlo procedures.

We will assume the existence of a table of successive random fractions R_f that are uniformly distributed (i.e. rectangular distribution) between 0 and 1. While physical tables of random numbers exist, the random fraction will normally be supplied as a standard function on a digital computer.

The distribution of the variate x may be prescribed by a normalized distribution function such that the probability of a value of x lying between x and $x + \mathrm{d}x$ is given by

$$f_x \,\mathrm{d}x. \tag{D1}$$

If the range of x is from a to b, then the total probability is

$$\int_a^b f_x \,\mathrm{d}x = 1. \tag{D2}$$

Now define the cumulative distribution function as

$$F_x = \int_a^x f_x \,\mathrm{d}x. \tag{D3}$$

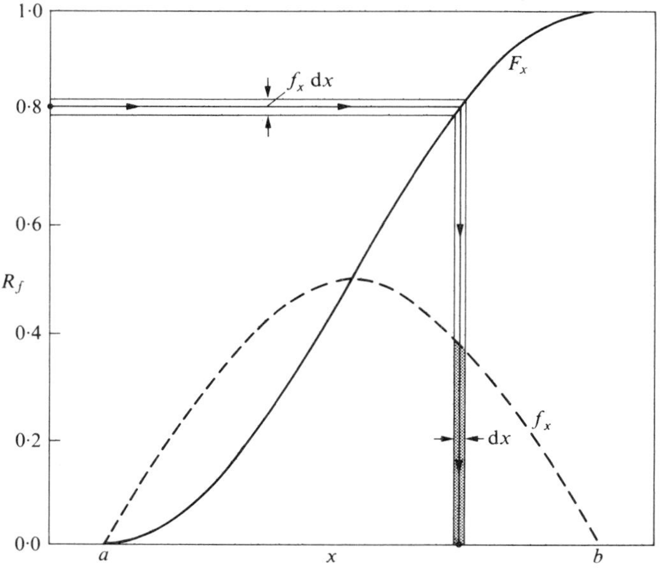

FIG. D.1. Typical normalized and cumulative distribution functions.

We may now generate a random fraction R_f and set this equal to F_x. The representative value of x is therefore given by

$$F_x = R_f. \tag{D4}$$

First consider the trivial example in which the variate x is uniformly distributed between a and b. Then

$$f_x = \text{const.}$$

and, using eqn (D2),

$$f_x = 1/(b-a).$$

Therefore, from eqn (D3),

$$F_x = \int_a^x \frac{1}{b-a}\,\mathrm{d}x = \frac{x-a}{b-a}.$$

and eqn (D4) finally gives

$$\frac{x-a}{b-a} = R_f,$$

or

$$x = a + R_f(b-a). \tag{D5}$$

This result is otherwise obvious from elementary considerations.

Now consider the variate r distributed between 0 and a such that the probability of r is proportional to r. This is the distribution that one would encounter in setting a random radius in a uniform axi-symmetric flow. Now

$$f_r = \text{const.} \times r$$

and, again evaluating the constant by eqn (D2),

$$f_r = 2r/a^2.$$

Therefore, from eqn (D3),

$$F_r = r^2/a^2$$

and eqn (D4) now gives

$$r = a(R_f)^{\frac{1}{2}}. \tag{D6}$$

Unfortunately, the above procedure can only be used when it is possible to invert eqn (D4) to obtain an explicit function for x. Consider the distribution function $f_{u'}$ for a thermal velocity component in an equilibrium gas. This is given by eqn (4.13) as

$$f_{u'} = (\beta/\pi^{\frac{1}{2}}) \exp(-\beta^2 u'^2) \tag{D7}$$

Therefore

$$F_{u'} = \frac{\beta}{\pi^{\frac{1}{2}}} \int_{-\infty}^{u'} \exp(-\beta^2 u'^2)\, du',$$

or

$$F_{u'} = \tfrac{1}{2}\{1 + \operatorname{erf}(\beta u')\}.$$

This expression cannot be inverted to give u' in terms of $R_{f'}$, and the method fails.

The general alternative is to use the *acceptance–rejection method*. In order to make direct use of the random fraction, the distribution function is divided by its maximum value $f_{\max}$ to give

$$f'_x = f_x/f_{\max}. \tag{D8}$$

A value of x is then chosen at random on the assumption that it is uniformly distributed between its limits. The function f'_x is then calculated for this value of x and a further random fraction R_f is generated. The value of x is then either accepted or rejected depending on whether f'_x is greater or less than this R_f. The procedure is repeated until a value of x is accepted. Since R_f is uniformly distributed between 0 and 1, the probability of a value of x being accepted is obviously proportional to f'_x and the successive accepted values conform to this distribution.

As an example, consider the function defined by eqn (D7). Its maximum is $\beta/\pi^{\frac{1}{2}}$ at $u' = 0$, so that $f'_{u'} = \exp(-\beta^2 u'^2)$. The uniformly distributed value of u' is given by eqn (D5) with a and b set as arbitrary finite cast-off values in place of the real limits of $-\infty$ to $+\infty$. If a and b are set at $-3/\beta$ and $+3/\beta$ respectively, the fraction of values lying outside these limits as $1-\text{erf}(3)$ or 0·000022 (see eqn (4.10)). Therefore,

$$u' = (-3+6R_f)/\beta$$

and

$$f'_{u'} = \exp\{-(-3+6R_f)^2\}.$$

The *next* value of R_f is then generated and, if $f'_{u'} > R_f$, u' is accepted. If $f'_{u'} < R_f$ the value of u' is rejected and the process is repeated until a value is accepted.

While the acceptance–rejection method involves a repetitive process which may require the evaluation of a large number of functions and random numbers, it can be used with almost any distribution function and is readily incorporated into computer programs.

Additional methods are available for special cases. One of these provides a direct method of sampling pairs of values from the normal distribution of eqn (D7). These values may be denoted by u' and v' and, from eqn (D7),

$$\begin{aligned} f_{u'}\,du' f_{v'}\,dv' &= (\beta/\pi^{\frac{1}{2}})\exp(-\beta^2 u'^2)\,du'(\beta/\pi^{\frac{1}{2}})\exp(-\beta^2 v'^2)\,dv' \\ &= (\beta^2/\pi)\exp\{-\beta^2(u'^2+v'^2)\}\,du'\,dv'. \end{aligned}$$

Now set

$$u' = r\cos\theta$$

and

$$v' = r\sin\theta. \tag{D9}$$

Then, since the Jacobean

$$\frac{\partial(u',v')}{\partial(r,\theta)} = \begin{vmatrix} \dfrac{\partial u'}{\partial r} & \dfrac{\partial u'}{\partial \theta} \\ \dfrac{\partial v'}{\partial r} & \dfrac{\partial v'}{\partial \theta} \end{vmatrix} = \begin{vmatrix} \cos\theta & -r\sin\theta \\ \sin\theta & r\cos\theta \end{vmatrix} = r,$$

$$\begin{aligned} f_{u'}\,du' f_{v'}\,dv' &= (\beta^2/\pi)\exp(-\beta^2 r^2) r\,dr\,d\theta \\ &= \exp(-\beta^2 r^2)\,d(\beta^2 r^2)\,d\theta/(2\pi) \end{aligned}$$

The angle θ is uniformly distributed between 0 and 2π so that, from eqn (D5),

$$\theta = 2\pi R_f. \tag{D10}$$

The variable $\beta^2 r^2$ is distributed between 0 and ∞ and its distribution function

$$f_{\beta^2 r^2} = \exp(-\beta^2 r^2)$$

is also already in a normalized form. The cumulative distribution function is

$$F_{\beta^2 r^2} = 1 - \exp(-\beta^2 r^2)$$

and, noting that R_f and $1-R_f$ are equivalent functions, eqn (D4) gives,

$$r = \{-\ln(R_f)\}^{\frac{1}{2}}/\beta. \tag{D11}$$

A pair of values of r and θ may be sampled from eqns (D10) and (D11) using successive random fractions. The normally distributed values of u' and v' follow from eqn (D9) and provide typical values for a thermal velocity component in an equilibrium gas.

APPENDIX E

Listing of the program for the cylindrical tube flux problem

```
      PROGRAM TEMC (INPUT,OUTPUT,TAPE5=INPUT,TAPE6=OUTPUT)
C
C FREE MOLECULE FLUX THROUGH CYLINDRICAL TUBE--TEST PARTICLE MONTE CARLO PROGRAM
C
      WRITE (6,1)
    1 FORMAT (44H1FREE MOLECULE FLUX THROUGH CYLINDRICAL TUBE///)
C READ DATA
      READ (5,2) TL,NTT
    2 FORMAT (F10.5,I10)
      WRITE (6,3) TL,NTT
    3 FORMAT (39H FLUX RATIOS FOR LENGTH-RADIUS RATIO OF,F10.5,9H BASED
     1ON,I10,28H REPRESENTATIVE TRAJECTORIES//)
      ND=NT=0
C ND IS THE NUMBER PASSING DIRECTLY THROUGH, AND NT IS THE TOTAL NUMBER THROUGH
      DO 4 N=1,NTT
C LOOP 4 IS OVER THE NTT TRAJECTORIES
      RM=SQRT(RANF(0))
C RM IS THE RANDOM RADIUS OF ENTRY
      CALL DCEM(AL,AM,AN)
      X=AL*(RM*AM+SQRT(AM*AM+(1.-RM*RM)*AN*AN))/(AM*AM+AN*AN)
C X IS X CO-ORD. OF THE FIRST INTERSECTION WITH THE CYLINDER
      IF (X.GT.TL) GO TO 5
    6 CALL DCEM(AM,AL,AN)
      X=X+2.*AL*AM/(AM*AM+AN*AN)
C X IS NOW THE CO-ORD. OF A SUBSEQUENT INTERSECTION
      IF (X.LT.0.) GO TO 4
      IF (X.GT.TL) GO TO 8
      GO TO 6
    5 ND=ND+1
    8 NT=NT+1
    4 CONTINUE
      FD=FLOAT(ND)/NTT
      FT=FLOAT(NT)/NTT
      WRITE (6,9) FD,FT
    9 FORMAT (41H FRACTION PASSING THROUGH TUBE DIRECTLY =,F10.5,33H ,AN
     1D THE TOTAL NUMBER FRACTION =,F10.5////)
      STOP
      END


      SUBROUTINE DCEM(DCP,DC1,DC2)
C
C GENERATES DIRECTION COSINES OF TYPICAL MOLECULE IN EFFUSING OR EMITTED GAS
C
C DCP IS WITH THE DIRECTION OF EFFUSION, DC1 AND DC2 ARE WITH NORMAL DIRECTIONS
      DCP=SQRT(RANF(0))
      A=SQRT(1.-DCP*DCP)
      B=6.28318531*RANF(0)
      DC1=A*COS(B)
      DC2=A*SIN(B)
      RETURN
      END
```

```
FREE MOLECULE FLUX THROUGH CYLINDRICAL TUBE


FLUX RATIOS FOR LENGTH-RADIUS RATIO OF    1.00000 BASED ON       5000 REPRESENTATIVE TRAJECTORIES

FRACTION PASSING THROUGH TUBE DIRECTLY =     .37880 ,AND THE TOTAL NUMBER FRACTION =     .67180
```

APPENDIX F

Collision rate in Monte Carlo simulation

The number of molecules in a cell is N_m, so that the total number of possible collision pairs is

$$N_\mathrm{p} = N_m(N_m - 1)/2. \tag{F1}$$

Therefore, c_r has a set of possible values $c_{\mathrm{r},m}$, where $m = 1, \dots, N_\mathrm{p}$. Now, let $c_{\mathrm{r},0}$ be some fixed reference value of c_r which, to a particular time t, has a probability of occurrence P. Since the collision pairs are chosen with probability proportional to $c_\mathrm{r}^{(\eta-5)/(\eta-1)}$, the probability of occurrence of c_r is

$$P(c_{\mathrm{r},m}/c_{\mathrm{r},0})^{(\eta-5)/(\eta-1)}.$$

The total number of collisions to the time t is obtained by summing over all possible collision pairs, i.e.

$$\sum_{m=1}^{N_\mathrm{p}} P\left(\frac{c_{\mathrm{r},m}}{c_{\mathrm{r},0}}\right)^{(\eta-5)/(\eta-1)} = \left(\frac{P}{c_{\mathrm{r},0}^{(\eta-5)/(\eta-1)}}\right) \sum_{m=1}^{N_\mathrm{p}} c_{\mathrm{r},m}^{(\eta-5)/(\eta-1)} \tag{F2}$$

Eqn (F2) shows that the time will be

$$t = \sum_{m=1}^{N_\mathrm{p}} \left(\frac{2}{N_m}\right) \{\pi W_{0,m}^2 (\kappa/m_\mathrm{r})^{2/(\eta-1)} n c_{\mathrm{r},m}^{(\eta-5)/(\eta-1)}\}^{-1} P\left(\frac{c_{\mathrm{r},m}}{c_{\mathrm{r},0}}\right)^{(\eta-5)/(\eta-1)}$$

and, since $c_{\mathrm{r},m}$ disappears,

$$t = P(2/N_m)\{\pi n W_{0,m}^2 (\kappa/m_\mathrm{r})^{2/(\eta-1)} n c_{\mathrm{r},0}^{(\eta-5)/(\eta-1)}\}^{-1} N_\mathrm{p}. \tag{F3}$$

The total number of collisions per unit time is, therefore,

$$\left(\frac{N_m}{2}\right) \pi n W_{0,m}^2 (\kappa/m_\mathrm{r})^{2/(\eta-1)} \left(\frac{1}{N_\mathrm{p}}\right) \sum_{m=1}^{N_\mathrm{p}} c_{\mathrm{r},m}^{(\eta-5)/(\eta-1)}$$

or

$$\left(\frac{N_m}{2}\right) \pi n W_{0,m}^2 (\kappa/m_\mathrm{r})^{2/(\eta-1)} \overline{c_\mathrm{r}^{(\eta-5)/(\eta-1)}}.$$

Since each collision involves two molecules, the mean collision rate per molecule is

$$\nu = \pi n W_{0,m}^2 (\kappa/m_\mathrm{r})^{2/(\eta-1)} \overline{c_\mathrm{r}^{(\eta-5)/(\eta-1)}}. \tag{F4}$$

The total collision cross-section of inverse power law molecules is given by eqn (2.27) as

$$\sigma_\mathrm{T} = \pi W_{0,m}^2 \left(\frac{\kappa}{m_\mathrm{r} c_\mathrm{r}^2}\right)^{\frac{2}{\eta-1}}$$

Since $W_{0,m}$, κ, and m_r are constants, this result may be substituted into eqn (F4) to give

$$\nu = n\overline{\sigma_\mathrm{T} c_\mathrm{r}},$$

in agreement with eqn (1.6).

APPENDIX G

Listing of the direct simulation program for the Rayleigh program

```
      PROGRAM DRAY (INPUT,OUTPUT,TAPE5=INPUT,TAPE6=OUTPUT)
C
C MONTE CARLO SIMULATION OF THE RAYLEIGH PROBLEM
C
      DIMENSION P(4,2000),LCR(2000),C(3,20),IC(2,20),SC(7,20,5 ),SS(7,5
     1),VRC(3),OP(10)
C
C  P(4,M) CONTAINS INFORMATION ON UP TO M SIMULATED MOLECULES
C    P(1,N),P(2,N),P(3,N) ARE U,V AND W VELOCITY COMPONENTS OF MOLECULE N
C    P(4,N) IS Y CO-ORDINATE OF MOLECULE N
C  LCR(M)    CONTAINS THE M MOLECULE NUMBERS ARRANGED IN ORDER OF THEIR CELLS
C
C  C(3,L),IC(2,L) CONTAIN INFORMATION ON UP TO L CELLS
C    C(1,N) IS THE Y CO-ORDINATE OF CELL N
C    C(2,N) IS THE TIME IN CELL N
C    C(3,N) IS THE MAXIMUM RELATIVE VELOCITY IN COLLISIONS IN CELL N
C    IC(1,N) CONTAINS THE NUMBER OF MOLECULES IN CELL N
C    IC(2,N) CONTAINS (STARTING ADDRESS-1) FOR MOLECULES OF CELL N IN LCR(M)
C
C  SC(7,L,J)  CONTAINS SAMPLED INFORMATION ON UP TO L CELLS AT UP TO J TIMES
C    SC(1,N,I) NUMBER IN SAMPLE FOR CELL N AT TIME I
C    SC(2,N,I) , SCSC(3,N,I) ARE SUMS OF U AND V MOLECULAR VELOCITIES, RESP.
C    SC(4,N,I),SC(5,N,I),SC(6,N,I) ARE SUMS OF U*U,V*V AND W*W, RESPECTIVELY
C    SC(7,N,I) NUMBER OF COLLISIONS IN CELL N OVER TIME INTERVAL I
C
C  SS(7,J). CONTAINS SAMPLED INFORMATION AT SURFACE IN UP TO J TIME INTERVALS
C    SS(1,N) NUMBER IN SAMPLE OVER NTH. TIME INTERVAL
C    SS(2,N) AND SS(3,N) INCIDENT AND REFLECTED NORMAL MOMENTUM OVER INTERVAL
C    SS(4,N) AND SS(5,N) INCIDENT AND REFLECTED PARALLEL MOMENTUM
C    SS(6,N)  AND SS(7,N) INCIDENT AND REFLECTED ENERGY OVER NTH. TIME INTERVAL
C
C  VRC(3) CONTAINS THE X,Y, AND Z COMPS. OF THE RELATIVE VELOCITY IN A COLLISION
C  OP(10) CONTAINS OUTPUT RESULTS
C
      WRITE (6,1)
    1 FORMAT (52H1DIRECT MOLECULAR SIMULATION OF THE RAYLEIGH PROBLEM//)
C
C READ DATA (SEE DATA OUTPUT FOR MEANINGS OF SYMBOLS)
      READ (5,2) TW,UW,YM,NC,MC,DTM,NIS,NST,NSEC
    2 FORMAT (3F10.5,2I10,F10.5,3I6)
C
C DATA PRINTOUT
      WRITE (6,3) TW,UW
    3 FORMAT (55H THE PLANE Y=0 IS DIFFUSELY REFLECTING WITH TEMPERATURE
     1,F10.5,35H AND VELOCITY IN THE X DIRECTION OF,F10.5)
      WRITE (6,4) YM
    4 FORMAT (28H THE FLOWFIELD EXTENDS TO Y=,F10.5,53H WHICH IS TAKEN T
     1O BE A SPECULARLY REFLECTING SURFACE)
      WRITE (6,5) NC,MC
    5 FORMAT (30H THE FLOWFIELD IS DIVIDED INTO,I6,41H UNIFORM CELLS, EA
     1CH INITIALLY CONTAINING,I6,10H MOLECULES)
      WRITE (6,6) DTM
    6 FORMAT (45H THE MOLECULES ARE MOVED AT TIME INTERVALS OF,F10.5)
      WRITE (6,7) NIS,NST
    7 FORMAT (26H PROPERTIES SAMPLED EVERY ,I6,30H INTERVALS AND RUN STO
     1PS AFTER,I6,10H SAMPLINGS)
      WRITE (6,8) NSEC
```

```
    8 FORMAT (1H ,I8,86H SECONDS OF COMPUTER TIME ARE AVAILABLE AND RUNS
     1 CONTINUE AS LONG AS TIME IS AVAILABLE///)
      WRITE (6,9)
    9 FORMAT (41H NORMALISATION OF VARIABLES IS AS FOLLOWS/)
      WRITE (6,10)
   10 FORMAT (75H A LENGTH IS LENGTH/L, WHERE L IS THE MEAN FREE PATH IN
     1 THE UNDISTURBED GAS)
      WRITE (6,11)
   11 FORMAT (100H A VELOCITY IS VELOCITY*B, WHERE B IS THE RECIPROCAL O
     1F THE MOST PROBABLE MOLECULAR SPEED IN THE GAS)
      WRITE (6,12)
   12 FORMAT (21H A TIME IS TIME/(B*L))
      WRITE (6,13)
   13 FORMAT (82H A TEMPERATURE IS TEMPERATURE/T, WHERE T IS THE TEMPERA
     1TURE OF THE UNDISTURBED GAS)
      WRITE (6,14)
   14 FORMAT (69H A DENSITY IS DENSITY/D, WHERE D IS THE DENSITY OF THE
     1FREESTREAM GAS)
      WRITE (6,15)
   15 FORMAT (81H A NUMBER FLUX IS FLUX*B/N, WHERE N IS THE NUMBER DENSI
     1TY OF THE UNDISTURBED GAS )
      WRITE (6,16)
   16 FORMAT  (59H A PRESSURE IS PRESSURE*B*B/D, SIMILARLY FOR SHEAR STR
     1ESSES)
      WRITE (6,17)
   17 FORMAT (33H A HEAT FLUX IS HEAT FLUX*B*B*B/D///)
C THE PLATE IS ASSUMED TO HAVE UNIT AREA
C
C SET CONSTANTS
      PI=3.141593
      CH=YM/NC
C CH IS CELL HEIGHT
      VMW=SQRT(TW)
C VMW IS THE MOST-PROBABLE MOLECULAR SPEED AT THE WALL TEMPERATURE
C (N.B. BOTH T AND B ARE UNITY IN THE UNDISTURBED GAS, THEREFORE GAS CONST.=0.5
      FND=MC/CH
C FND IS THE UN-NORMALISED UNDISTURBED GAS NUMBER DENSITY
      CXS=1./(SQRT(2.)*FND)
C CXS IS THE COLLISION CROSS SECTION OF THE HARD SPHERE MOLECULES
C  (SINCE MEAN FREE PATH =1 IN UNDISTURBED GAS OF NUMBER DENSITY FND)
      NRUN=0
C WORKING VARIABLES AND THOSE WITH OBVIOUS MEANINGS ARE LEFT UNDEFINED
      DO 21 N=1,NC
   21 C(1,N)=(N-0.5)*CH
      NM=NC*MC
C NM IS THE (CONSTANT) TOTAL NUMBER OF MOLECULES
      ACU=DTM*NIS*MC*FND*CXS*SQRT(2./PI)
C ACU IS THE NUMBER OF COLLISIONS PER CELL PER SAMPLING INTERVAL IN UNDIST. GAS
      VRM=2.*SQRT(2./PI)
C VRM IS THE AVERAGE RELATIVE SPEED IN UNDISTURBED GAS COLLISIONS
C
C SET SAMPLING VARIABLES TO ZERO
      DO 18 N=1,NST
      DO 18 M=1,7
      SS(M,N)=0
      DO 18 L=1,NC
   18 SC(M,L,N)=0
C
C START OF RUN
   20 NRUN=NRUN+1
      CALL SECOND(TSR)
C TSR IS THE COMPUTER TIME AT THE START OF THE RUN
C
C INITIAL STATE OF THE GAS WILL NOW BE SET
      DO 22 N=1,NC
      C(2,N)=RANF(0)*2.*CH/(MC*MC*CXS*VRM)
C INITIAL CELL TIME IS A RANDOM FRACTION OF THE MEAN COLLISION TIME INCREMENT
      C(3,N)=2.*VRM
C MAXIMUM RELATIVE VELOCITY WILL BE RESET IF FASTER ENCOUNTERS OCCUR
      DO 22 M=1,MC
      J=(N-1)*MC+M
      P(4,J)=C(1,N)+(RANF(0)-0.5)*CH
      DO 22 L=1,3
      B=SQRT(-ALOG(RANF(0)))
      A=2.*PI*RANF(0)
   22 P(L,J)=B*SIN(A)
      DO 23 J=1,NST
```

```
C
C LOOP OVER SAMPLING INTERVALS
      DO 24 I=1,NIS
C
C LOOP OVER THE TIME INTERVALS WITHIN SAMPLING INTERVAL J
      TIME=((J-1)*NIS+I)*DTM
C TIME IS THE OVERALL FLOW TIME
C
C MOLECULES NOW MOVE APPROPRIATELY TO TIME INTERVAL DTM
      DO 25 N=1,NM
      Y=P(4,N)+P(2,N)*DTM
      IF (Y.LT.YM) GO TO 26
C
C MOLECULE REFLECTS FROM SPECULARLY REFLECTING PLANE AT Y=YM
      Y=1.9999999*YM-Y
      P(2,N)=-P(2,N)
   26 IF (Y.GT.0.) GO TO 25
C
C MOLECULE REFLECTS FROM WALL AT Y=0
      DTR=DTM*Y/(Y-P(4,N))
      IF (DTR.LT.1.E-10) DTR=1.E-10
C DTR IS TIME REMAINING AFTER MOLECULE STRIKES WALL
      P(1,N)=P(1,N)-UW
C THIS CHANGES FRAME OF REFERENCE FOR SAMPLING WALL PROPERTIES
      SS(1,J)=SS(1,J)+1.
      SS(2,J)=SS(2,J)-P(2,N)
      SS(4,J)=SS(4,J)+P(1,N)
      SS(6,J)=SS(6,J)+0.5*(P(1,N)*P(1,N)+P(2,N)*P(2,N)+P(3,N)*P(3,N))
C INCIDENT MOLECULE SAMPLING HAS BEEN SET (MASS OF MOLECULE IS UNITY)
      B=VMW*SQRT(-ALOG(RANF(0)))
      A=2.*PI*RANF(0)
      P(1,N)=B*SIN(A)
      P(3,N)=B*COS(A)
      P(2,N)=VMW*SQRT(-ALOG(RANF(0)))
      Y=P(2,N)*DTR
      SS(3,J)=SS(3,J)+P(2,N)
      SS(5,J)=SS(5,J)+P(1,N)
      SS(7,J)=SS(7,J)+0.5*(P(1,N)*P(1,N)+P(2,N)*P(2,N)+P(3,N)*P(3,N))
      P(1,N)=P(1,N)+UW
C REFLECTED MOLECULE PROPERTIES AND SAMPLING HAVE NOW BEEN SET
   25 P(4,N)=Y
C
C MOLECULAR INDEXING WILL NOW BE SET
      DO 27 N=1,NC
   27 IC(1,N)=0
      DO 28 N=1,NM
      M=P(4,N)/CH+0.9999999
      IF (M.EQ.0) M=1
C ABOVE STATEMENT IS TO GUARD AGAINST ROUND-OFF ERROR
   28 IC(1,M)=IC(1,M)+1
      M=0
      DO 29 N=1,NC
      IC(2,N)=M
      M=M+IC(1,N)
   29 IC(1,N)=0
      DO 30 N=1,NM
      M=P(4,N)/CH+0.9999999
      IF (M.EQ.0) M=1
      IC(1,M)=IC(1,M)+1
      K=IC(2,M)+IC(1,M)
   30 LCR(K)=N
C
C CALCULATE COLLISIONS
      DO 24 N=1,NC
      IF (C(2,N).GT.TIME) GO TO 24
      IF (IC(1,N).GE.2) GO TO 32
      C(2,N)=C(2,N)+DTM
C COLLISIONS BYPASSED IF CELL TIME EXCEEDS TIME OR IF ONLY ONE MOL. IN CELL
      GO TO 24
   32 K=RANF(0)*IC(1,N)+IC(2,N)+0.9999999
      IF (K.EQ.IC(2,N)) K=K+1
      L=LCR(K)
   33 K=RANF(0)*IC(1,N)+IC(2,N)+0.9999999
      IF (K.EQ.IC(2,N)) K=K+1
      M=LCR(K)
      IF (M.EQ.L) GO TO 33
```

```
C TWO MOLECULES HAVE BEEN SELECTED AT RANDOM FROM CELL N
      DO 34 K=1,3
   34 VRC(K)=P(K,L)-P(K,M)
      VR=SQRT(VRC(1)*VRC(1)+VRC(2)*VRC(2)+VRC(3)*VRC(3))
C VR IS THE RELATIVE SPEED OF THE MOLECULES
      IF (VR.GT.C(3,N)) C(3,N)=VR
      A=VR/C(3,N)
      B=RANF(0)
      IF (A.LT.B) GO TO 32
C PROBABILITY OF COLLISION OF PAIR IS PROPORTIONAL TO VR FOR HARD SPHERE MOLS.
      C(2,N)=C(2,N)+2.*CH/(IC(1,N)*IC(1,N)*CXS*VR)
C THE CELL TIME HAS BEEN ADVANCED BY THE AMOUNT APPROPRIATE TO THE COLLISION
      SC(7,N,J)=SC(7,N,J)+1.
      B=1.-2.*RANF(0)
      A=SQRT(1.-B*B)
      VRC(1)=B*VR
      B=2.*PI*RANF(0)
      VRC(2)=A*COS(B)*VR
      VRC(3)=A*SIN(B)*VR
C VRC(3) NOW CONTAINS THE COMPONENTS OF THE POST COLLISION RELATIVE VELOCITY
      DO 36 K=1,3
      VCCM=0.5*(P(K,L)+P(K,M))
C VCCM IS THE APPROPRIATE COMPONENT OF THE CENTRE OF  MASS VEL. OF COLL. PAIR
      P(K,L)=VCCM+VRC(K)*0.5
   36 P(K,M)=VCCM-VRC(K)*0.5
      IF (C(2,N).LT.TIME) GO TO 32
   24 CONTINUE
C
C FLOW FIELD WILL NOW BE SAMPLED
      DO 23 N=1,NC
      K=IC(1,N)
      DO 23 M=1,K
      LL=IC(2,N)+M
      L=LCR(LL)
      SC(1,N,J)=SC(1,N,J)+1.
      SC(2,N,J)=SC(2,N,J)+P(1,L)
      SC(3,N,J)=SC(3,N,J)+P(2,L)
      DO 23 KK=1,3
   23 SC(KK+3,N,J)=SC(KK+3,N,J)+P(KK,L)*P(KK,L)
C CHECK WHETHER THERE IS TIME FOR A FURTHER RUN
      CALL SECOND (TER)
      TFR=TER-TSR
C TER IS COMPUTER TIME AT END OF RUN AND TFR IS COMPUTER TIME FOR RUN
      IF (NSEC-TER.GT.1.5*TFR) GO TO 20
C
C NO TIME FOR FURTHER RUN, SO PRINT RESULTS
      WRITE (6,31)
   31 FORMAT (19H SURFACE PROPERTIES///)
      WRITE (6,37)
   37 FORMAT (95H    TIME INTERVAL       SAMPLE    NUMBER          PRESSURE
     1          SHEAR STRESS           HEAT FLUX)
      WRITE (6,38)
   38 FORMAT(100H    FROM        TO         SIZE       FLUX    INCIDENT   REFL
     1ECTED INCIDENT  REFLECTED INCIDENT  REFLECTED/)
      DO 39 N=1,NST
      OP(1)=(N-1)*DTM*NIS
      OP(2)=N*DTM*NIS
      OP(3)=SS(1,N)
      A=NRUN*NIS*DTM*FND
      DO 40 M=4,10
   40 OP(M)=SS(M-3,N)/A
   39 WRITE (6,41) OP
   41 FORMAT (1H ,2F 9.5,F11.0,7F10.5)
      DO 42 J=1,NST
      A=J*DTM*NIS
      WRITE (6,43) A
   43 FORMAT (32HOFLOW FIELD PROPERTIES AT TIME =,F10.5//)
      WRITE (6,44)
   44 FORMAT(112H CELL  Y CO-ORD   SAMPLE   DENSITY    X-VEL.    Y-VEL.
     1  X-TEMP.   Y-TEMP.   Z-TEMP.      TEMP.   COLL. RATE CELL/)
      DO 45 N=1,NC
      OP(1)=SC(1,N,J)
      OP(2)=OP(1)/(NRUN*MC)
      OP(3)=SC(2,N,J)/OP(1)
      OP(4)=SC(3,N,J)/OP(1)
```

```
      OP(5)=2.*(SC(4,N,J)/OP(1)-OP(3)*OP(3))
      OP(6)=2.*(SC(5,N,J)/OP(1)-OP(4)*OP(4))
      OP(7)=2.*SC(6,N,J)/OP(1)
      OP(8)=(OP(5)+OP(6)+OP(7))/3.
      OP(9)=SC(7,N,J)/(NRUN*ACU)
   45 WRITE (6,46) N,C(1,N),(OP(L),L=1,9),N
   46 FORMAT (1H ,I4,F10.5,F 9.0,8F10.5,I6)
   42 CONTINUE
      STOP
      END
```

DIRECT MOLECULAR SIMULATION OF THE RAYLEIGH PROBLEM

THE PLANE Y=0 IS DIFFUSELY REFLECTING WITH TEMPERATURE 1.60000 AND VELOCITY IN THE X DIRECTION OF 2.00000
THE FLOWFIELD EXTENDS TO Y= 10.00000 WHICH IS TAKEN TO BE A SPECULARLY REFLECTING SURFACE
THE FLOWFIELD IS DIVIDED INTO 20 UNIFORM CELLS, EACH INITIALLY CONTAINING 50 MOLECULES
THE MOLECULES ARE MOVED AT TIME INTERVALS OF .17725
PROPERTIES SAMPLED EVERY 5 INTERVALS AND RUN STOPS AFTER 5 SAMPLINGS
1024 SECONDS OF COMPUTER TIME ARE AVAILABLE AND RUNS CONTINUE AS LONG AS TIME IS AVAILABLE

NORMALISATION OF VARIABLES IS AS FOLLOWS

A LENGTH IS LENGTH/L, WHERE L IS THE MEAN FREE PATH IN THE UNDISTURBED GAS
A VELOCITY IS VELOCITY*B, WHERE B IS THE RECIPROCAL OF THE MOST PROBABLE MOLECULAR SPEED IN THE GAS
A TIME IS TIME/(B*L)
A TEMPERATURE IS TEMPERATURE/T, WHERE T IS THE TEMPERATURE OF THE UNDISTURBED GAS
A DENSITY IS DENSITY/D, WHERE D IS THE DENSITY OF THE FREESTREAM GAS
A NUMBER FLUX IS FLUX*B/N, WHERE N IS THE NUMBER DENSITY OF THE UNDISTURBED GAS
A PRESSURE IS PRESSURE*B*B/D, SIMILARLY FOR SHEAR STRESSES
A HEAT FLUX IS HEAT FLUX*B*B*B/D

SURFACE PROPERTIES

TIME INTERVAL FROM	TO	SAMPLE SIZE	NUMBER FLUX	PRESSURE INCIDENT	PRESSURE REFLECTED	SHEAR STRESS INCIDENT	SHEAR STRESS REFLECTED	HEAT FLUX INCIDENT	HEAT FLUX REFLECTED
0.00000	.88625	3145.	.29087	.26760	.32640	-.54455	-.00641	.84429	.46733
.88625	1.77250	3249.	.30049	.29019	.33068	-.48888	-.00585	.81393	.46597
1.77250	2.65875	3172.	.29337	.29971	.33169	-.41645	.00414	.73886	.47182
2.65875	3.54500	3168.	.29300	.30057	.32964	-.39144	-.00272	.72101	.46741
3.54500	4.43125	3177.	.29383	.31812	.33402	-.35552	-.00019	.70284	.47805

FLOW FIELD PROPERTIES AT TIME = .88625

CELL	Y CO-ORD	SAMPLE	DENSITY	X-VEL.	Y-VEL.	X-TEMP.	Y-TEMP.	Z-TEMP.	TEMP.	COLL. RATE	CELL
1	.25000	5764.	.94492	.67167	.02047	2.52524	1.39062	1.46316	1.79301	1.16882	1
2	.75000	6196.	1.01574	.23172	.05788	1.60391	1.24999	1.21647	1.35679	1.10128	2
3	1.25000	6217.	1.01918	.06553	.05041	1.22076	1.17482	1.12234	1.17264	1.03833	3
4	1.75000	6236.	1.02230	.02844	.02252	1.08292	1.05685	1.03437	1.05805	1.03014	4
5	2.25000	6128.	1.00459	.01090	.00740	1.03041	1.00488	.99224	1.00918	1.02456	5
6	2.75000	5957.	.97656	-.01055	.00418	.98280	1.00162	1.00399	.99614	.97079	6
7	3.25000	6153.	1.00869	-.01115	-.00309	.97430	.96659	1.00993	.98364	1.02522	7
8	3.75000	6124.	1.00393	.00567	.01370	.96894	.98101	1.00251	.98415	1.01866	8
9	4.25000	6089.	.99820	.01108	-.00814	.99924	1.01428	1.00462	1.00605	.99899	9
10	4.75000	6168.	1.01115	-.01573	-.00002	1.01907	1.01624	.97648	1.00393	1.04096	10
11	5.25000	6120.	1.00328	.00102	.00030	1.01768	.98754	1.00129	1.00217	1.00883	11
12	5.75000	6017.	.98639	-.02263	-.00511	1.01181	.97208	1.00431	.99606	1.01440	12
13	6.25000	6132.	1.00525	.00109	.01353	1.00301	.99344	1.02207	1.00617	1.01309	13
14	6.75000	6073.	.99557	-.00128	.00645	1.01358	.97451	.99737	.99515	.99670	14
15	7.25000	6124.	1.00393	.01886	-.01464	.99870	.99600	.98129	.99199	.98915	15
16	7.75000	5972.	.97902	.00044	.00105	1.02086	1.00624	1.00788	1.01166	1.03145	16
17	8.25000	6144.	1.00721	.00360	.00875	1.00071	.98653	1.00443	.99722	1.01309	17
18	8.75000	6213.	1.01852	.00453	.00251	1.01369	1.04786	1.01752	1.02636	1.05571	18
19	9.25000	6148.	1.00787	-.00055	.00372	.97722	1.00166	1.02633	1.00174	1.01079	19
20	9.75000	6025.	.98770	-.00805	.01244	.99539	.95794	1.00998	.98777	.99506	20

FLOW FIELD PROPERTIES AT TIME = 1.77250

CELL	Y CO-ORD	SAMPLF	DENSITY	X-VEL.	Y-VEL.	X-TEMP.	Y-TEMP.	Z-TEMP.	TEMP.	COLL. RATE	CELL
1	.25000	5453.	.89393	.91529	.00438	2.50500	1.47507	1.59916	1.85974	1.10161	1
2	.75000	5785.	.94836	.52163	.03352	2.06787	1.37420	1.48817	1.64342	1.19440	2
3	1.25000	6137.	1.00607	.27272	.07034	1.57304	1.26967	1.27075	1.37115	1.19767	3
4	1.75000	6429.	1.05393	.14154	.09697	1.30410	1.24279	1.20948	1.25212	1.19013	4
5	2.25000	6349.	1.04082	.05429	.08372	1.17124	1.20206	1.12367	1.16565	1.11440	5
6	2.75000	6252.	1.02492	.02216	.05065	1.05385	1.15459	1.06704	1.09183	1.03604	6
7	3.25000	6177.	1.01262	.01197	.01916	1.03255	1.08115	1.02227	1.04532	1.02227	7
8	3.75000	6266.	1.02721	-.00998	.00458	1.01248	1.03155	.98521	1.00975	1.05407	8
9	4.25000	6072.	.99541	-.01448	-.00974	1.00563	1.05149	.97803	1.01172	1.01538	9
10	4.75000	6047.	.99131	.00473	.01521	.99006	1.00839	.97275	.99040	1.01571	10
11	5.25000	6109.	1.00148	.00493	.00683	1.01406	.99580	1.01678	1.00888	1.02784	11
12	5.75000	6057.	.99295	-.01304	.00932	1.03474	1.00346	1.02829	1.02216	1.02817	12
13	6.25000	6181.	1.01328	-.00795	.00935	1.00671	1.00905	.96961	.99512	.99276	13
14	6.75000	6192.	1.01508	.00155	-.00314	.98036	.97105	1.00690	.98610	1.02391	14
15	7.25000	6001.	.98377	.00567	-.01125	1.03657	1.02788	.98546	1.01664	1.00391	15
16	7.75000	6073.	.99557	.00766	-.00371	1.03463	.98497	.99625	1.00528	1.00391	16
17	8.25000	6026.	.98787	-.00602	-.00877	1.01433	.99012	1.01128	1.00524	1.01047	17
18	8.75000	6043.	.99066	.02750	.01486	.99282	1.01493	1.01170	1.00648	1.02128	18
19	9.25000	6092.	.99869	-.00293	.01596	.98136	.99279	1.03980	1.00465	.99014	19
20	9.75000	6259.	1.02607	-.01460	-.00628	.97445	1.00118	.99116	.98893	1.02686	20

FLOW FIELD PROPERTIES AT TIME = 2.65875

CELL	Y CO-ORD	SAMPLE	DENSITY	X-VEL.	Y-VEL.	X-TEMP.	Y-TEMP.	Z-TEMP.	TEMP.	COLL. RATE	CELL
1	.25000	5239.	.85885	1.04309	-.00189	2.51536	1.44565	1.68018	1.88039	1.02883	1
2	.75000	5497.	.90115	.68188	.02991	2.30111	1.49714	1.50042	1.76622	1.10751	2
3	1.25000	5887.	.96508	.45349	.05424	1.83251	1.38823	1.45407	1.55827	1.15833	3
4	1.75000	6092.	.99869	.28193	.06891	1.53497	1.31842	1.32836	1.39392	1.20292	4
5	2.25000	6333.	1.03820	.15554	.08620	1.32702	1.25133	1.22949	1.26928	1.22882	5
6	2.75000	6576.	1.07803	.08959	.10215	1.24403	1.23241	1.15367	1.21004	1.24128	6
7	3.25000	6603.	1.08246	.03669	.08728	1.16083	1.22223	1.09521	1.15942	1.15735	7
8	3.75000	6304.	1.03344	.01531	.07459	1.05343	1.15437	1.05944	1.08908	1.11440	8
9	4.25000	6200.	1.01639	-.00567	.03676	1.05230	1.12416	1.01254	1.06633	1.04948	9
10	4.75000	6003.	.98410	.03669	.02786	1.02274	1.09097	1.03990	1.05120	.99276	10
11	5.25000	6214.	1.01869	.01242	.01288	1.03280	1.06501	1.02471	1.04094	1.03997	11
12	5.75000	6150.	1.00820	-.00840	.00574	1.01154	1.03995	.98216	1.01121	1.02489	12
13	6.25000	6148.	1.00787	-.00155	-.00046	1.04042	1.01397	1.02609	1.02683	1.05309	13
14	6.75000	6289.	1.03098	.00158	.01052	.98574	1.02642	.98328	.99848	1.06325	14
15	7.25000	6060.	.99344	-.00720	.00528	.99285	1.03255	.98057	1.00199	.97899	15
16	7.75000	5972.	.97902	.01095	.00972	1.00241	.99756	1.00898	1.00298	1.00686	16
17	8.25000	5940.	.97377	.00425	.01263	1.02217	.97494	1.00789	1.00166	.96588	17
18	8.75000	6110.	1.00164	.00596	.00090	.98824	1.01117	1.03296	1.01079	.97702	18
19	9.25000	6214.	1.01869	-.00686	-.01485	1.02013	.98373	1.01142	1.00509	1.05735	19
20	9.75000	6169.	1.01131	.00046	.00286	1.00372	.96555	.98831	.98586	1.02817	20

FLOW FIELD PROPERTIES AT TIME = 3.54500

CELL	Y CO-ORD	SAMPLE	DENSITY	X-VEL.	Y-VEL.	X-TEMP.	Y-TEMP.	Z-TEMP.	TEMP.	COLL. RATE	CELL
1	.25000	5157.	.84541	1.16828	-.00436	2.46437	1.58249	1.71143	1.91943	1.02751	1
2	.75000	5307.	.87000	.80976	.01865	2.28659	1.48251	1.65733	1.80881	1.02948	2
3	1.25000	5582.	.91508	.57612	.02270	1.88091	1.40240	1.52607	1.60313	1.08096	3
4	1.75000	5848.	.95869	.38433	.06340	1.65344	1.34700	1.46620	1.48888	1.15505	4
5	2.25000	6083.	.99721	.24821	.06789	1.50920	1.29159	1.31110	1.37063	1.19144	5
6	2.75000	6469.	1.06049	.15997	.09775	1.33698	1.26959	1.29121	1.29926	1.31111	6
7	3.25000	6594.	1.08098	.09941	.10512	1.25072	1.25975	1.20095	1.23714	1.28357	7
8	3.75000	6569.	1.07689	.04993	.11136	1.15830	1.21007	1.15278	1.17705	1.21374	8
9	4.25000	6407.	1.05033	.02525	.09241	1.12708	1.23504	1.06780	1.14331	1.16456	9
10	4.75000	6396.	1.04852	.02982	.09024	1.09971	1.19698	1.10301	1.13323	1.10653	10
11	5.25000	6269.	1.02770	.00695	.04965	1.05915	1.14356	1.05273	1.08515	1.07735	11
12	5.75000	6257.	1.02574	.00088	.03285	1.05121	1.08044	1.02841	1.05335	1.05866	12
13	6.25000	6155.	1.00902	.00949	.03209	1.02843	1.04756	1.03477	1.03692	1.06522	13
14	6.75000	6162.	1.01016	-.01171	.02255	1.01691	1.02076	1.00060	1.01276	1.03604	14
15	7.25000	6136.	1.00590	.00056	.02815	1.01900	1.05594	1.00580	1.02691	1.01178	15
16	7.75000	6062.	.99377	.00129	.00484	.98143	.98791	1.03745	1.00226	1.00588	16
17	8.25000	6086.	.99770	.01277	.00083	1.01595	1.04126	1.00912	1.02211	1.01604	17
18	8.75000	6164.	1.01049	-.00228	.01463	1.00427	1.00313	.99727	1.00156	1.00883	18
19	9.25000	6106.	1.00098	-.00471	-.00933	1.00241	.98486	1.01433	1.00053	1.02555	19
20	9.75000	6191.	1.01492	.00694	.00377	.98616	.98370	1.02964	.99984	1.03866	20

FLOW FIELD PROPERTIES AT TIME = 4.43125

CELL	Y CO-ORD	SAMPLE	DENSITY	X-VEL.	Y-VEL.	X-TEMP.	Y-TEMP.	Z-TEMP.	TEMP.	COLL. RATE	CELL
1	.25000	5028.	.82426	1.24205	.00859	2.42509	1.61646	1.75386	1.93140	.99456	1
2	.75000	5229.	.85721	.90803	.00916	2.26315	1.51779	1.65149	1.81081	1.03784	2
3	1.25000	5446.	.89279	.69417	.04418	2.00254	1.50126	1.60883	1.70421	1.04456	3
4	1.75000	5624.	.92197	.50079	.05593	1.80724	1.43305	1.50080	1.59036	1.08718	4
5	2.25000	5835.	.95164	.32826	.05357	1.58275	1.34122	1.40353	1.44250	1.14981	5
6	2.75000	6251.	1.02475	.22927	.08632	1.41729	1.31405	1.33864	1.35666	1.25210	6
7	3.25000	6366.	1.04361	.16159	.09918	1.31919	1.23089	1.28199	1.27736	1.27111	7
8	3.75000	6560.	1.07541	.08885	.10060	1.25353	1.24117	1.19639	1.23036	1.28652	8
9	4.25000	6546.	1.07311	.06080	.11378	1.21995	1.22676	1.16947	1.20539	1.26587	9
10	4.75000	6515.	1.06803	.05011	.12386	1.17620	1.20236	1.14018	1.17291	1.22128	10
11	5.25000	6508.	1.06689	.02240	.10215	1.10258	1.15809	1.08862	1.11643	1.19013	11
12	5.75000	6407.	1.05033	.00326	.08310	1.07896	1.15463	1.08731	1.10696	1.14358	12
13	6.25000	6325.	1.03689	.01038	.07861	1.07540	1.14116	1.08521	1.10059	1.11112	13
14	6.75000	6328.	1.03738	.00526	.05330	1.04949	1.10807	1.04778	1.06845	1.07964	14
15	7.25000	6337.	1.03885	-.00396	.02347	1.03505	1.08124	1.03481	1.05037	1.07604	15
16	7.75000	6112.	1.00197	.00576	.03271	1.02401	1.04398	1.03219	1.03339	1.05014	16
17	8.25000	6035.	.98934	.01504	.01153	1.01526	1.06116	1.00313	1.02652	.97735	17
18	8.75000	6215.	1.01885	-.00342	-.00005	1.02634	1.04572	.99371	1.02192	1.06850	18
19	9.25000	6198.	1.01607	-.00055	.01226	1.02093	1.03775	.98999	1.01622	1.01604	19
20	9.75000	6165.	1.01066	.00425	-.00944	.99714	1.01361	1.00556	1.00543	1.06686	20

APPENDIX H

Routine for handling variations in the number of simulated molecules

```
PROGRAM DEMR(INPUT,OUTPUT,TAPE5=INPUT,TAPE6=OUTPUT)
C
C  ROUTINE FOR HANDLING VARIATIONS IN THE NUMBER OF MOLECULES
C
C  THIS IS NOT A COMPLETE PROGRAM
C THE ARRAY P(5,M) CONTAINS THE FLOATING POINT INFORMATION ON MOLECULE M, AND
C           IP(3,M) CONTAINS THE INTEGER INFORMATION ON THIS MOLECULE
C
      N=0
C N IS THE MOLECULE COUNTING INTEGER
    1 N=N+1
      IF (N.GT.NM) GO TO 2
C NM IS THE TOTAL NUMBER OF MOLECULES, AND LABEL 2 MARKS THE END OF THE ROUTINE
      IF (IP(1,N).LT.0) GO TO 1
C THIS IS A SECONDARY APPLICATION OF THE INTEGER IP(1,N), AND THE NEGATIVE VALUE
C     TAGS MOLECULES THAT HAVE BEEN ADDED IN THIS ROUTINE AS ARESULT OF THE
C     DUPLICATION PROCEDURE. IP(1,N) IS ALWAYS POSITIVE (E.G. THE CELL NUMBER) IN
C     ITS PRIMARY APPLICATION AND IS NOT USED AGAIN BEFORE BEING RESET.
C
C A NUMBER OF STATEMENTS MUST BE INCLUDED AT THIS POINT TO
C
C (A)  MOVE MOLECULE N OVER THE TIME INTERVAL DTM TO A NEW LOCATION
C (B)  IF IT MOVES OUTSIDE THE SIMULATED REGION, TRANSFER TO LABEL 3
C (C)  OTHERWISE IT MOVES FROM CELL J TO CELL K (J MAY BE EQUAL TO K)
C (D)  IF WEIGHTING FACTORS ARE NOT USED, TRANSFER TO LABEL 1
C
      IF (J.EQ.K) GO TO 1
      B=W(J)/W(K)
C W(I) IS THE WEIGHTING FACTOR IN CELL I
      M=0
    4 IF (B.LT.1.) GO TO 5
      M=M+1
      B=B-1
      GO TO 4
    5 A=RANF(0)
C RANF(0) GENERATES A RANDOM FRACTION BETWEEN 0 AND 1
      IF (A.LT.B) M=M+1
      IF (M.EQ.0) GO TO 3
C THE MOLECULE IS DISCARDED IF M=0
      IF (M.EQ.1) GO TO 1
C NO FURTHER ACTION IS REQUIRED IF M=1
      M=M-1
      DO 6 L=1,M
C THE MOLECULE IS DUPLICATED IN THE LOOP OVER LABEL 6
      NM=NM+1
      DO 7 I=1,5
    7 P(I,NM)=P(I,N)
      DO 8 I=1,3
    8 IP(I,NM)=IP(I,N)
    6 IP(1,NM)=-1
      GO TO 1
    3 DO 9 I=1,5
    9 P(I,N)=P(I,NM)
      DO 10 I=1,3
   10 IP(I,N)=IP(I,NM)
C MOLECULE N HAS BEEN REMOVED THROUGH ITS REPLACEMENT BY MOLECULE NM
      NM=NM-1
      N=N-1
      GO TO 1
    2 CONTINUE
```

APPENDIX I

Simulation program for the relaxation of a homogeneous gas mixture

```
      PROGRAM RMIX (INPUT,OUTPUT,TAPE5=INPUT,TAPE6=OUTPUT)
C
C TEMPERATURE RELAXATION IN A HARD SPHERE GAS MIXTURE
C
      DIMENSION P(3,4000,3),C(3,6),TC(3,3),SC(3,3),T(3,3),VM(3,3),VRC(3)
C
C P(L,M,N) CONTAINS INFORMATION ON MOLECULE M OF GAS SPECIES L
C   N=1,2,AND 3 FOR X,Y,AND Z VELOCITY COMPONENTS, RESPECTIVELY
C C(L,N) CONTAINS INFORMATION ON GAS SPECIES L
C   N=1 NUMBER DENSITY
C   N=2 MOLECULAR MASS
C   N=3 MOLECULAR DIAMETER
C   N=4 INITIAL TEMPERATURE
C   N=5 WEIGHTING FACTOR
C   N=6 SUM OF SQUARES OF VELOCITIES
C TC(L,M) IS THE THEORETICAL EQUILIBRIUM COLLISION RATE PER MOLECULE OF SPECIES
C         L WITH ONE OF SPECIES M
C SC(L,M) COUNTS THE ACTUAL COLLISIONS PER MOLECULE OF SPECIES L MOLECULES WITH
C         SPECIES M MOLECULES
C T(L,M) IS THE TIME PARAMETER FOR SPECIES L-M COLLISIONS
C VM(L,M) IS THE EXPECTED MAXIMUM RELATIVE VELOCITY FOR SPECIES L-M COLLISIONS
C VRC(3) CONTAINS THE X,Y, AND Z COMPONENTS OF THE RELATIVE VELOCITY
      WRITE (6,1)
    1 FORMAT (38H TEMPERATURE RELAXATION IN GAS MIXTURE///)
C
C READ DATA
      READ (5,2) NC,NM,NP,DTP
    2 FORMAT (3I10,F10.5)
      WRITE (6,3) NC,NM
    3 FORMAT (1H ,I4,34H  GAS SPECIES AND A SAMPLE SIZE OF,I7,9H FOR EAC
     1H)
      WRITE (6,4) NP,DTP
    4 FORMAT (1H ,I4,36H  PRINTING INTERVALS, EACH EQUAL TO ,F10.5,57H T
     1IMES THE MEAN COLLISION TIME IN THE EQUILIBRIUM MIXTURE/)
      DO 5 L=1,NC
      READ (5,6) (C(L,N),N=1,4)
    6 FORMAT (4F10.5)
    5 WRITE (6,7)  L,(C(L,N),N=1,4)
    7 FORMAT (8H SPECIES,I4,19H OF NUMBER DENSITY ,F10.5,7H ,MASS ,F10.5
     1,11H ,DIAMETER ,F10.5,27H , AND INITIAL TEMPERATURE ,F10.5/)
C THE NORMALIZATION IS SUCH THAT THE RESULTS DEPEND ONLY ON THE RATIOS RATHER
C THAN THE ABSOLUTE VALUES OF THE MOLECULAR PROPERTIES
C THE BOLTZMANN CONSTANT IS CONVENIENTLY SET EQUAL TO 0.5
C
C SET CONSTANTS
      PI=3.14159265
      SPI=SQRT(PI)
      SN=SNT=0.
      DO 8 L=1,NC
      SN=SN+C(L,1)
      SNT=SNT+C(L,1)*C(L,4)
C SN IS THE TOTAL NUMBER DENSITY AND SNT IS A MEASURE OF THE TOTAL ENERGY
      C(L,3)=C(L,3)/C(1,3)
      C(L,5)=C(L,1)
C THERE IS AN EQUAL NUMBER OF SIMULATED MOLECULES OF EACH SPECIES ,THE NUMBER
C DENSITY THEREFORE BECOMES THE WEIGHTING FACTOR
    8 C(L,6)=0.
    ANM=NM
    TFE=SNT/SN
```

```
C TFE IS THE FINAL EQUILIBRIUM TEMPERATURE
C EQN (4.41) MAY BE USED TO CALCULATE THE MOLECULAR DIAMETERS SUCH THAT THE MEAN
C COLLISION RATE WILL BE UNITY IN THE FINAL EQUILIBRIUM STATE
      A=0.
      DO 9 L=1,NC
      DO 9 M=1,NC
    9 A=A+(C(L,1)/SN)*(SPI/2.)*(C(L,3)+C(M,3))**2*C(M,1)*SQRT(TFE*(C(L,2
     1)+C(M,2))/(C(L,2)*C(M,2)))
      C(1,3)=1./SQRT(A)
      DO 10 N=2,NC
   10 C(N,3)=C(1,3)*C(N,3)
      WRITE (6,38) TFE
   38 FORMAT (32H FINAL EQUILIBRIUM TEMPERATURE =,F10.5,49H WITH A MEAN
     1COLLISION RATE PER MOLECULE OF UNITY)
      DO 11 L=1,NC
      DO 36 M=1,NC
      TC(L,M)=(SPI/2.)*(C(L,3)+C(M,3))**2*C(M,1)*SQRT(TFE*(C(L,2)+C(M,2)
     1)/(C(L,2)*C(M,2)))
      SC(L,M)=0.
      T(L,M)=0.
   36 VM(L,M)=2.*(SQRT(C(L,2)*C(L,4))+SQRT(C(M,2)*C(M,4)))
      WRITE (6,37) L
   37 FORMAT (69H THE THEORETICAL EQUILIBRIUM COLLISION RATE FOR MOLECUL
     1ES OF SPECIES ,I4,36H WITH SPECIES 1 ,SPECIES 2  ETC IS -)
   11 WRITE (6,30) (TC(L,M),M=1,NC)
C
C SET INITIAL STATE OF MOLECULES
      DO 12 L=1,NC
      D=SQRT(C(L,4)/C(L,2))
      DO 12 M=1,NM
      DO 12 N=1,3
   13 V=-3.+6.*RANF(0)
      A=EXP(-V*V)
      B=RANF(0)
      IF (A.LT.B ) GO TO 13
   12 P(L,M,N) =V*D
      DO 14 NT=1,NP
      TIME =NT*DTP
C
C CALCULATE COLLISIONS
      DO 15 L=1,NC
      DO 15 M=1,NC
   22 IF (T(L,M).GT.TIME) GO TO 15
   16 K=RANF(0)*NM+0.999999
      IF (K.EQ.0) K=K+1
   17 J=RANF(0)*NM+0.999999
      IF (J.EQ.0) J=J+1
      IF (L.EQ.M.AND.J.EQ.K) GO TO 17
      DO 18 N=1,3
   18 VRC(N)=P(L,K,N)-P(M,J,N)
      VR=SQRT(VRC(1)*VRC(1)+VRC(2)*VRC(2)+VRC(3)*VRC(3))
      IF (VM(L,M).LT.VR) VM(L,M)=VR
      A=VR/VM(L,M)
      B=RANF(0)
      IF (A.LT.B) GO TO 16
C A SPECIES L-M PAIR HAS NOW BEEN SELECTED
      LP=MP=1
      B=RANF(0)
      A=C(L,5)/C(M,5)
      IF (A.GT.1.) GO TO 19
      IF (A.LT.B) MP=0
      GO TO 20
   19 A=1./A
      IF (A.LT.B) LP=0
C LP AND MP ARE THE PROBABILITIES THAT THE COLLISION WILL BE COUNTED FOR THE
C          L AND M SPECIES, RESPECTIVELY
   20 CXS=PI*(C(L,3)+C(M,3))**2/4.
      T(L,M)=T(L,M)+FLOAT(LP)/(CXS*C(M,1)*VR*NM)+FLOAT(MP)/(CXS*C(L,1)*V
     1R*NM)
      SC(L,M)=SC(L,M)+FLOAT(LP)/ANM/DTP
      SC(M,L)=SC(M,L)+FLOAT(MP)/ANM/DTP
      B=1.-2.*RANF(0)
      A=SQRT(1.-B*B)
      VRC(1)=B*VR
      B=2.*PI*RANF(0)
      VRC(2)=A*COS(B)*VR
      VRC(3)=A*SIN(B)*VR
      SM=C(L,2)+C(M,2)
```

```
      RML=C(L,2)/SM
      RMM=C(M,2)/SM
      DO 21 N=1,3
      VCCM=RML*P(L,K,N)+RMM*P(M,J,N)
      IF (LP.EQ.1) P(L,K,N)=VCCM+VRC(N)*RMM
      IF (MP.EQ.1) P(M,J,N)=VCCM-VRC(N)*RML
   21 CONTINUE
      GO TO 22
   15 CONTINUE
C
C SAMPLE TEMPERATURES AND PRINT RESULTS
      WRITE (6,23)
   23 FORMAT (1H ///)
      WRITE (6,24) TIME
   24 FORMAT(16H RESULTS AT TIME,F10.5/)
      A=0.
      DO 25 L=1,NC
      DO 25 M=1,NC
   25 A=A+SC(L,M)*C(L,1)/SN
      WRITE (6,26) A
   26 FORMAT (61H MEAN COLLISION RATE PER MOLECULE OVER PREVIOUS TIME IN
     1TERVAL,F10.5)
      WRITE (6,27)
   27 FORMAT (45H INDIVIDUAL COMPONENTS OF THIS COLLISION RATE)
      DO 28 L=1,NC
      WRITE (6,29) L
   29 FORMAT (8H SPECIES,I4,46H MOLECULE WITH SPECIES 1 , SPECIES 2 AND
     1SO ON)
   28 WRITE (6,30) (SC(L,M),M=1,NC)
   30 FORMAT (1H ,4F12.5)
      DO 31 L=1,NC
      DO 31 M=1,NM
      DO 31 N=1,3
   31 C(L,6)=C(L,6)+P(L,M,N)*P(L,M,N)
      TF=0.
      DO 32 L=1,NC
   32 TF=TF+C(L,1)*C(L,2)*C(L,6)
      TF=0.6666667*TF/(ANM*SN)
      WRITE (6,33) TF
   33 FORMAT (22H OVERALL TEMPERATURE =,F10.5)
      DO 34 L=1,NC
      A=0.6666667*C(L,2)*C(L,6)/ANM
      WRITE(6,35)L,A
   35 FORMAT (23H TEMPERATURE OF SPECIES,I4,3H = ,F10.5)
      C(L,6)=0.
      DO 34 M=1,NC
   34 SC(L,M)=0.
   14 CONTINUE
      STOP
      END
```

```
TEMPERATURE RELAXATION IN GAS MIXTURE


   3  GAS SPECIES AND A SAMPLE SIZE OF    4000 FOR EACH
  60  PRINTING INTERVALS, EACH EQUAL TO      .50000 TIMES THE MEAN COLLISION TIME IN THE EQUILIBRIUM MIXTURE

SPECIES   1 OF NUMBER DENSITY    1.00000 ,MASS    1.00000 ,DIAMETER    1.00000 , AND INITIAL TEMPERATURE    9.00000

SPECIES   2 OF NUMBER DENSITY    1.00000 ,MASS    3.00000 ,DIAMETER    2.00000 , AND INITIAL TEMPERATURE    5.00000

SPECIES   3 OF NUMBER DENSITY    1.00000 ,MASS   10.00000 ,DIAMETER     .50000 , AND INITIAL TEMPERATURE    1.00000

FINAL EQUILIBRIUM TEMPERATURE =   5.00000 WITH A MEAN COLLISION RATE PER MOLECULE OF UNITY
THE THEORETICAL EQUILIBRIUM COLLISION RATE FOR MOLECULES OF SPECIES    1 WITH SPECIES 1 ,SPECIES 2  ETC IS -
      .32080      .58935      .13383
THE THEORETICAL EQUILIBRIUM COLLISION RATE FOR MOLECULES OF SPECIES    2 WITH SPECIES 1 ,SPECIES 2  ETC IS -
      .58935      .74086      .23332
THE THEORETICAL EQUILIBRIUM COLLISION RATE FOR MOLECULES OF SPECIES    3 WITH SPECIES 1 ,SPECIES 2  ETC IS -
      .13383      .23332      .02536



RESULTS AT TIME    .50000

MEAN COLLISION RATE PER MOLECULE OVER PREVIOUS TIME INTERVAL   1.15867
INDIVIDUAL COMPONENTS OF THIS COLLISION RATE
SPECIES   1 MOLECULE WITH SPECIES 1 , SPECIES 2 AND SO ON
      .42700      .74000      .17500
SPECIES   2 MOLECULE WITH SPECIES 1 , SPECIES 2 AND SO ON
      .74000      .77700      .21400
SPECIES   3 MOLECULE WITH SPECIES 1 , SPECIES 2 AND SO ON
      .17500      .21400      .01400
OVERALL TEMPERATURE =   5.01350
TEMPERATURE OF SPECIES   1 =    8.27481
TEMPERATURE OF SPECIES   2 =    5.37832
TEMPERATURE OF SPECIES   3 =    1.38737
```

```
RESULTS AT TIME    1.00000

MEAN COLLISION RATE PER MOLECULE OVER PREVIOUS TIME INTERVAL    1.14567
INDIVIDUAL COMPONENTS OF THIS COLLISION RATE
SPECIES   1 MOLECULE WITH SPECIES 1 , SPECIES 2 AND SO ON
      .41700      .71100      .16900
SPECIES   2 MOLECULE WITH SPECIES 1 , SPECIES 2 AND SO ON
      .71100      .79600      .22450
SPECIES   3 MOLECULE WITH SPECIES 1 , SPECIES 2 AND SO ON
      .16900      .22450      .01500
OVERALL TEMPERATURE =    5.01350
TEMPERATURE OF SPECIES   1 =     7.72958
TEMPERATURE OF SPECIES   2 =     5.62152
TEMPERATURE OF SPECIES   3 =     1.68939
```

....

```
RESULTS AT TIME   29.50000

MEAN COLLISION RATE PER MOLECULE OVER PREVIOUS TIME INTERVAL     .98500
INDIVIDUAL COMPONENTS OF THIS COLLISION RATE
SPECIES   1 MOLECULE WITH SPECIES 1 , SPECIES 2 AND SO ON
      .29700      .58850      .13300
SPECIES   2 MOLECULE WITH SPECIES 1 , SPECIES 2 AND SO ON
      .58850      .72800      .22950
SPECIES   3 MOLECULE WITH SPECIES 1 , SPECIES 2 AND SO ON
      .13300      .22950      .02800
OVERALL TEMPERATURE =    5.01350
TEMPERATURE OF SPECIES   1 =     5.00418
TEMPERATURE OF SPECIES   2 =     4.99715
TEMPERATURE OF SPECIES   3 =     5.03916


RESULTS AT TIME   30.00000


MEAN COLLISION RATE PER MOLECULE OVER PREVIOUS TIME INTERVAL     .99333
INDIVIDUAL COMPONENTS OF THIS COLLISION RATE
SPECIES   1 MOLECULE WITH SPECIES 1 , SPECIES 2 AND SO ON
      .33100      .57600      .13950
SPECIES   2 MOLECULE WITH SPECIES 1 , SPECIES 2 AND SO ON
      .57600      .73600      .22750
SPECIES   3 MOLECULE WITH SPECIES 1 , SPECIES 2 AND SO ON
      .13950      .22750      .02700
OVERALL TEMPERATURE =    5.01350
TEMPERATURE OF SPECIES   1 =     5.04310
TEMPERATURE OF SPECIES   2 =     4.96910
TEMPERATURE OF SPECIES   3 =     5.02829
```

APPENDIX J

Simulation program for the relaxation of internal degrees of freedom in a homogeneous gas

```
      PROGRAM RINT (INPUT,OUTPUT,TAPE5=INPUT,TAPE6=OUTPUT)
C
C INTERNAL TEMPERATURE RELAXATION IN A HOMOGENEOUS GAS WITH ENERGY SINK MODEL
C
      DIMENSION  P(4,4000),VRC(3)
C
C P(4,M) CONTAINS INFORMATION ON UP TO M SIMULATED MOLECULES
C   P(1,N),P(2,N),P(3,N) ARE THE U,V, AND W VELOCITY COMPONENTS OF MOLECULE N
C   P(4,N) IS THE INTERNAL ENERGY OF MOLECULE N
C VRC(3) CONTAINS THE X,Y, AND Z COMPONENTS OF THE RELATIVE VEL. IN A COLLISION
C VCM(3) CONTAINS THE X,Y, AND Z COMPONENTS OF THE CENTRE OF MASS VELOCITY
C NORMALIZATION OF TEMPERATURE IS TO THE INITIAL TRANSLATIONAL TEMPERATURE
C NORMALIZATION OF TIME IS TO THE PRODUCT OF THE MEAN FREE PATH AND THE INVERSE
C OF THE MOST PROBABLE MOLECULAR SPEED IN THE INITIAL GAS
      WRITE (6,1)
    1 FORMAT (32H1INTERNAL TEMPERATURE RELAXATION///)
C
C READ DATA
      READ (5,2) NM,KTM,DFI,TFAC,TIN,DTP,NT
    2 FORMAT (2I10,4F10.5,I10)
      WRITE (6,3) NM,DFI
    3 FORMAT (1H ,I6,18H MOLECULES HAVING ,F10.5,28H INTERNAL DEGREES OF
     1 FREEDOM)
      IF (KTM.EQ.0) WRITE (6,4)
      IF (KTM.EQ.1) WRITE (6,5)
    4 FORMAT (22H HARD SPHERE MOLECULES)
    5 FORMAT (29H INVERSE 9TH. POWER MOLECULES)
      WRITE (6,6) TFAC
    6 FORMAT (18H TRANSFER FACTOR =,F10.5)
      WRITE (6,7) TIN
    7 FORMAT (57H INITIAL RATIO OF INTERNAL TO TRANSLATIONAL TEMPERATURE
     1 =,F10.5)
      WRITE (6,8) NT,DTP
    8 FORMAT (1H ,I4,28H PRINTING INTERVALS, EACH OF,F10.5//)
C
C SET CONSTANTS
      PI=3.141592654
      IF (KTM.EQ.0) CXS=SQRT(0.5)
      IF (KTM.EQ.1) CXS=2.57
C CXS IS THE COLLISION CROSS SECTION FOR HARD SPHERE MOLECULES OR THE EXPRESSION
C     OF EQN (8.17) FOR INVERSE NINTH POWER MOLECULES
      IF (KTM.EQ.0) BN=2.
      IF (KTM.EQ.1) BN=1.75
C BN IS THE COEFFICIENT OF KT IN THE MEAN TRANSLATIONAL RELATIVE ENERGY
      TC=0.
C TC IS THE TIME COUNTER FOR COLLISIONS
      NCOL=0
C NCOL IS THE NUMBER OF COLLISIONS
      IF (KTM.EQ.0) VRM=2.
      IF (KTM.EQ.1) VRM=1.5
C VRM IS THE MAXIMUM RELATIVE VELOCITY (OR ITS SQUARE ROOT FOR INVERSE NINTH
C     POWER MOLECULES) IN A COLLISION  (IT IS RESET AS NECESSARY)
C
C SET INITIAL STATE OF GAS
      DO 9 M=1,NM
      DO 10 N=1,3
   11 V=-3.+6.*RANF(0)
      A=EXP(-V*V)
      B=RANF(0)
      IF (A.LT.B) GO TO 11
```

```
   10 P(N,M)=V
    9 P(4,M)=(DFI/4.)*TIN
      DO 33 K=1,NT
      TIME=K*DTP
C
C CALCULATE COLLISIONS
   12 IF (TC.GT.TIME) GO TO 13
   14 J=RANF(0)*NM+0.999999
      IF (J.EQ.0) J=1
   15 L=RANF(0)*NM+0.999999
      IF (L.EQ.0) L=1
      IF (L.EQ.J) GO TO 15
      VRR=0.
      DO 16 N=1,3
      VRC(N)=P(N,J)-P(N,L)
   16 VRR=VRR+VRC(N)*VRC(N)
      VR=SQRT(VRR)
      IF (KTM.EQ.0) VRE=VR
      IF (KTM.EQ.1) VRE=SQRT(VR)
C PROBABILITY OF COLLISION IS PROPORTIONAL TO VRE
      IF (VRE.GT.VRM) VRM=VRE
      A=VRE/VRM
      B=RANF(0)
      IF (A.LT.B) GO TO 14
C SELECTION OF COLLISION PAIR HAS NOW BEEN MADE WITH APPROPRIATE PROBABILITY
      TC=TC+(2./FLOAT(NM))/(CXS*VRE)
      NCOL=NCOL+1
C TIME AND COLLISION COUNTERS HAVE NOW BEEN ADVANCED
      A=VRR*DFI*0.125/BN
C A IS THE AVERAGE EQUILIBRIUM MOLECULAR ENERGY APPROPRIATE TO THIS COLLISION
      DERJ=TFAC*(P(4,J)-A)
      DERL=TFAC*(P(4,L)-A)
      P(4,J)=P(4,J)-DERJ
      P(4,L)=P(4,L)-DERL
      VRR=VRR+(DERL+DERJ)*4.
      VRF=SQRT(VRR)/VR
C ENERGY HAS NOW BEEN TRANSFERRED BETWEEN THE TRANSLATIONAL AND INTERNAL MODES
C VRF IS THE FRACTIONAL CHANGE IN THE MAGNITUDE OF THE POST COLLISION RELATIVE
C      VELOCITY AS A CONSEQUENCE OF THIS TRANSFER
      IF (KTM.EQ.0) GO TO 17
C
C INVERSE NINTH POWER COLLISION MECHANICS
      WA=SQRT(RANF(0))*1.5
      EPS=2.*PI*RANF(0)
C IMPACT PARAMETERS HAVE BEEN CHOSEN WITH APROPRIATE PROBABILITY
      A=WA*(1.26233+WA*(1.84145+WA*(-8.87881+WA*(20.3313+WA*(-23.8155+WA
     1*(14.5046+WA*(-4.42027+WA*0.535193)))))))
      CHI=PI-2.*A
      CC=COS(CHI)
      SC=SIN(CHI)
      CE=COS(EPS)
      SE=SIN(EPS)
      DU=VRC(1)
      DV=VRC(2)
      DW=VRC(3)
      A=SQRT(DV*DV+DW*DW)
      VRC(1)=(DU*CC+SC*SE*A)*VRF
      VRC(2)=(DV*CC+SC*(VR*DW*CE-DU*DV*SE)/A)*VRF
      VRC(3)=(DW*CC-SC*(VR*DV*CE+DU*DW*SE)/A)*VRF
      GO TO 18
C
C HARD SPHERE COLLISION MECHANICS
   17 B=1.-2.*RANF(0)
      A=SQRT(1.-B*B)
      VRC(1)=B*VR*VRF
      B=2.*PI*RANF(0)
      VRC(2)=A*COS(B)*VR*VRF
      VRC(3)=A*SIN(B)*VR*VRF
C VRC(3) NOW CONTAINS THE POST COLLISION RELATIVE VELOCITY COMPONENTS
   18 DO 19 N=1,3
      VCCM=0.5*(P(N,J)+P(N,L))
      P(N,J)=VCCM+VRC(N)*0.5
   19 P(N,L)=VCCM-VRC(N)*0.5
      GO TO 12
   13 WRITE (6,20) TIME
```

```
C
C SAMPLE AND PRINT TEMPERATURES
   20 FORMAT (8H TIME = ,F10.5)
      WRITE(6,21) NCOL
   21 FORMAT (1H ,I6,11H COLLISIONS)
      A=B=0.
      DO 22 N=1,NM
      A=A+P(1,N)*P(1,N)+P(2,N)*P(2,N)+P(3,N)*P(3,N)
   22 B=B+P(4,N)
      TTR=0.6666667*A/NM
      TROT=4.*B/(DFI*NM)
      TOV=(3.*TTR+DFI*TROT)/(3.+DFI)
   33 WRITE (6,23) TOV,TTR,TROT
   23 FORMAT (25H TEMPERATURES, OVERALL = ,F10.5,17H TRANSLATIONAL = ,F1
     10.5,14H ROTATIONAL = ,F10.5//)
      STOP
      END
```

```
INTERNAL TEMPERATURE RELAXATION


  4000 MOLECULES HAVING     2.00000 INTERNAL DEGREES OF FREEDOM
HARD SPHERE MOLECULES
TRANSFER FACTOR =     .10000
INITIAL RATIO OF INTERNAL TO TRANSLATIONAL TEMPERATURE =    0.00000
  80 PRINTING INTERVALS, EACH OF     .50000


TIME =      .50000
  1111 COLLISIONS
TEMPERATURES, OVERALL =     .59731 TRANSLATIONAL =     .96110 ROTATIONAL =     .05164


TIME =     1.00000
  2193 COLLISIONS
TEMPERATURES, OVERALL =     .59731 TRANSLATIONAL =     .92959 ROTATIONAL =     .09889


TIME =     1.50000
  3275 COLLISIONS
TEMPERATURES, OVERALL =     .59731 TRANSLATIONAL =     .90169 ROTATIONAL =     .14075


TIME =     2.00000
  4314 COLLISIONS
TEMPERATURES, OVERALL =     .59731 TRANSLATIONAL =     .87721 ROTATIONAL =     .17747


TIME =     2.50000
  5380 COLLISIONS
TEMPERATURES, OVERALL =     .59731 TRANSLATIONAL =     .85409 ROTATIONAL =     .21215


TIME =     3.00000
  6423 COLLISIONS
TEMPERATURES, OVERALL =     .59731 TRANSLATIONAL =     .83356 ROTATIONAL =     .24294


TIME =     3.50000
  7455 COLLISIONS
TEMPERATURES, OVERALL =     .59731 TRANSLATIONAL =     .81436 ROTATIONAL =     .27175


TIME =     4.00000
  8484 COLLISIONS
TEMPERATURES, OVERALL =     .59731 TRANSLATIONAL =     .79681 ROTATIONAL =     .29807


TIME =     4.50000
  9469 COLLISIONS
TEMPERATURES, OVERALL =     .59731 TRANSLATIONAL =     .78187 ROTATIONAL =     .32047


TIME =     5.00000
 10452 COLLISIONS
TEMPERATURES, OVERALL =     .59731 TRANSLATIONAL =     .76895 ROTATIONAL =     .33986
```

```
TIME =    35.50000
 65526 COLLISIONS
TEMPERATURES, OVERALL =     .59731 TRANSLATIONAL =     .60396 ROTATIONAL =     .58734

TIME =    36.00000
 66406 COLLISIONS
TEMPERATURES, OVERALL =     .59731 TRANSLATIONAL =     .60356 ROTATIONAL =     .58795

TIME =    36.50000
 67247 COLLISIONS
TEMPERATURES, OVERALL =     .59731 TRANSLATIONAL =     .60375 ROTATIONAL =     .58767

TIME =    37.00000
 68154 COLLISIONS
TEMPERATURES, OVERALL =     .59731 TRANSLATIONAL =     .60300 ROTATIONAL =     .58879

TIME =    37.50000
 69033 COLLISIONS

TEMPERATURES, OVERALL =     .59731 TRANSLATIONAL =     .60286 ROTATIONAL =     .58899

TIME =    38.00000
 69909 COLLISIONS
TEMPERATURES, OVERALL =     .59731 TRANSLATIONAL =     .60277 ROTATIONAL =     .58913

TIME =    38.50000
 70808 COLLISIONS
TEMPERATURES, OVERALL =     .59731 TRANSLATIONAL =     .60236 ROTATIONAL =     .58975

TIME =    39.00000
 71675 COLLISIONS
TEMPERATURES, OVERALL =     .59731 TRANSLATIONAL =     .60315 ROTATIONAL =     .58855

TIME =    39.50000
 72560 COLLISIONS
TEMPERATURES, OVERALL =     .59731 TRANSLATIONAL =     .60312 ROTATIONAL =     .58860

TIME =    40.00000
 73456 COLLISIONS
TEMPERATURES, OVERALL =     .59731 TRANSLATIONAL =     .60314 ROTATIONAL =     .58857
```

REFERENCES

ABRAMOWITZ, M. and STEGUN, I. A. (1965). *Handbook of mathematical functions*. Dover, New York.

ALDER, B. J. and WAINWRIGHT, T. (1958). In *Proceedings of the international symposium on transport processes in statisical mechanics*, p. 97. Interscience, New York.

BARANTSEV, R. G. (1972). In *Progress in aerospace sciences*, Vol. 13 (ed. D. Kuchemann), p. 1. Pergamon Press, Oxford.

BHATNAGAR, P. L., GROSS, E. P., and KROOK, M. (1954). *Phys. Rev.* **94**, 511.

BIRD, G. A. (1963). *Phys. Fluids*, **6**, 1518.

—— (1969). In *Rarefied gas dynamics* (eds. L. Trilling and H. Y. Wachman), p. 301. Academic Press, New York.

—— (1970a). *Phys. Fluids*, **13**, 1172.

—— (1970*b*). *AIAA J.*, **8**, 1998.

—— (1970*c*). Seventh international symposium on rarefied gas dynamics, Pisa.

—— (1973). In *Proc. AIAA computational fluid dynamics conference*, p. 103. American Institute of Aeronautics and Astronatutics, New York.

BOLTZMANN, L. (1872). *Sber. Akad. Wiss. Wien Abt. II*, **66**, 275.

BROADWELL, J. E. (1964). *J. Fluid Mech.*, **19**, 401.

BRYAN, G. H. (1894). *Rep. Br. Ass. Advant. Sci.*, p. **83**.

CAMAC, M. (1965). In *Rarefied gas dynamics* (ed. J. H. de Leeuw), p. 240. Academic Press, New York.

CATTOLICA, R. ROBBEN, F., TALBOT, L., and WILLIS, D. R. (1974). *Phys. Fluids*, **17**, 1793.

CERCIGNANI, C. (1969). *Mathematical methods in kinetic theory*. Plenum Press, New York.

—— (1974). In *Rarefied gas dynamics* (ed. K. Karamcheti), p. 55. Academic Press, New York.

—— and LAMPIS, M. (1974). In *Rarefied gas dynamics* (ed. K. Karamcheti), p. 361. Academic Press, New York.

CHAPMAN, S. and COWLING, T. G. (1952). *The mathematical theory of non-uniform gases* (2nd edn). Cambridge University Press.

—— and —— (1970). *The mathematical theory of non-uniform gases* (3rd edn). Cambridge University Press.

CHU, C. K. (1967). In *Rarefied gas dynamics* (ed. C. L. Brundin), p. 589. Academic Press, New York.

CLAUSING, P. (1932). *Annln. Phys.* **12**, 961.

CRAWFORD, D. R. and VOGENITZ, F. W. (1974). In *Rarefied gas dynamics* (eds. M. Becker and M. Fiebig), p. B. 24. DFVLR Press, Porz-Wahn, Federal Republic of Germany.

CURTISS, C. F. and MUCKENFUSS, C. (1958), *J. chem. Phys.*, **29**, 1257.

DAHLER, J. S. and SATHER, N. F. (1962). *J. chem. Phys.*, **38**, 2363.

DAVIS, D. H. (1960). *J. appl. Phys.*, **31**, 1169.

DERZKO, N. A. (1972). Review of Monte Carlo methods in kinetic theory. *UTIAS Rev.* 35, University of Toronto.

DESPHANDE, S. M. and NARASIMHA, R. (1969). *J. Fluid Mech.*, **36**, 545.

DIEWERT, G. S. (1973). *Phys. Fluids*, **16**, 1215.

EDWARDS, R. H. and CHENG, H. K. (1966). *AIAA J.*, **4**, 558.

EPSTEIN, M. (1967). *AIAA J.*, **5**, 1797.

FIZSDON, W., HERCZYNSKI, R., and WALENTA, Z. (1974). In *Rarefied gas dynamics* (eds. M. Becker and M. Fiebig), p. Ax B23. DFVLR Press, Porz-Wahn, Federal Republic of Germany.

GIEDT, R. D., COHEN, N., and JACOBS, T. A. (1969). *J. chem. Phys.*, **50**, 5374.

GILBARG, D. and PAOLUCCI, D. (1953). *J. rat. Mech. Anal.*, **2**, 617.

GOODMAN, F. O. (1968). *Surface Sci.*, **11**, 283.

—— (1971). *Surface Sci.*, **26**, 327.

GRAD, H. (1949). *Commun. pure appl. Math.*, **2**, 331.

—— (1958). In *Encyclopaedia of Physics* (ed. S. Flügge) vol. XII, p. 205. Springer-Verlag, Berlin.

—— and HU, P. N. (1969). In *Rarefied gas dynamics* (eds. L. Trilling and H. Y. Wachman), p. 561. Academic Press, New York.

HAMEL, B. B. and WILLIS, D. R. (1966). *Phys. Fluids*, **9**, 829.

HARRIS, S. (1971). *An introduction to the theory of the Boltzmann equation*. Holt, Rinehart, and Winston, New York.

HAVILAND, J. K. (1965). In *Methods of computational physics* (ed. B. Alder *et al.*), **4**, p. 109. Academic Press, New York.

HICKS, B. L., YEN, S. M., and REILLY, B. L. (1972). *J. Fluid Mech.*, **53**, 85.

HINSHELWOOD, C. N. (1940). *The kinetics of chemical change*, p. 39. Clarendon Press, Oxford.

HIRSCHFELDER, J. O., CURTISS, C. F., and BIRD, R. B. (1954). *The molecular theory of gases and liquids*. Wiley, New York.

HOLTZ, T. (1974). Measurements of molecular velocity distribution functions in an argon normal shock wave at Mach number 7. Ph.D. Dissertation, University of Southern California.

JEANS, J. H. (1904). *Q. Jl. pure appl. Math.*, **25**, 224.

KENNARD, E. H. (1938). *Kinetic theory of gases*. McGraw-Hill, New York.

KOURA, K. (1973). *J. chem. Phys.*, **59**, 691.

KUSCER, J. (1971). *Surface Sci.*, **25**, 225.

—— (1974). In *Rarefied gas dynamics* (eds. M. Becker and M. Fiebig). p. El. DFVLR Press, Porz-Wahn, Federal Republic of Germany.

LARSEN, P. S. and BORGNAKKE, C. (1973). DCAMM Report 52. The Technical University of Denmark.

—— and —— (1974). In *Rarefied gas dynamics* (eds. M. Becker and M. Fiebig), p. A7. DFVLR Press, Porz-Wahn, Federal Republic of Germany.

LIEPMANN, H. W. (1961). *J. fluid Mech.*, **10**, 65.

—— and ROSHKO, A. (1957). *Elements of gas dynamics*. Wiley, New York.

——, NARASIMHA, R., and CHAHINE, M. T. (1962). *Phys. Fluids*, **5**, 1313.

LINZER, M. and HORNIG, D. F. (1963). *Phys. Fluids*, **6**, 1661.

LIU, C. Y. and LEES, L. (1961). In *Rarefied gas dynamics* (ed. L. Talbot), p. 391. Academic Press, New York.

LORDI, J. A. and MATES, R. E. (1970). *Phys. Fluids*. **13**, 291.

MAXWELL, J. C. (1867). *Phil. Trans. Soc.*, **157**, 49.

—— (1879). *Phil. Trans. R. Soc.* 1, Appendix.

MELVILLE, W. K. (1972). *J. Fluid Mech.*, **51**, 571.

METCALF, S. C., LILLICRAP, D. C., and BERRY, C. J. (1969). In *Rarefied gas dynamics* (eds. L. Trilling and H. Y. Wachman), p. 619. Academic Press, New York.

MILLER, D. R. and SUBBARAO, R. B. (1971). *J. chem. Phys.*, **55**, 1478.

MORDUCHOW, M. and LIBBY, P. A. (1949). *J. aeronaut. Sci.*, **16**, 11.

MOTT-SMITH, H. M. (1951). *Phys. Rev.*, **82**, 885.
MUCKENFUSS, C. (1962). *Phys. Fluids*, **5**, 1325.
MUNTZ, E. P. (1967). In *Rarefied gas dynamics* (ed. C. L. Brundin), p. 1257. Academic Press, New York.
NARASIMHA, R. (1962). *J. Fluid Mech.*, **12**, 294.
NOCILLA, S. (1963). In *Rarefied gas dynamics* (ed. J. A. Laurmann), p. 327. Academic Press, New York.
NORDSIECK, A. and HICKS, B. L. (1967). In *Rarefied gas dynamics* (ed. C. L. Brundin), p. 695. Academic Press, New York.
PATTERSON, G. N. (1971). *Introduction to the kinetic theory of gas flows*. University of Toronto Press, Toronto.
PIDDUCK, F. B. (1922). *Proc. R. Soc. A.*, **101**, 101.
ROBBEN, F. and TALBOT, L. (1966). *Phys. Fluids*, **9**, 633, 644, and 653.
RODE, D. L. and TANENBAUM, B. S. (1967). *Phys. Fluids*, **10**, 1352.
ROSS, J., LIGHT, J. C., and SCHULER, K. E. (1969). In *Kinetic processes in gases and plasmas* (ed. A. R. Hochstim), p. 281. Academic Press. New York.
RUSSELL, D. A. (1965). In *Rarefied gas dynamics* (ed. J. H. de Leeuw), p. 265. Academic Press, New York.
SANDLER, S. I. and DAHLER, J. S. (1967). *J. chem. Phys.*, **47**, 2621.
SATHER, N. F. (1973). *Phys. Fluids*, **16**, 2106.
SCHAAF, S. A. and CHAMBRE, P. L. (1961). *Flow of rarefied gases*. Princeton University Press.
SCHMIDT, B. (1969). *J. Fluid Mech.*, **39**, 361.
SCHULTZ-GRUNOW, F. and FROHN, A. (1965). In *Rarefied gas dynamics* (ed. J. H. De Leeuw), p. 250. Academic Press, New York.
SHERMAN, F. S. (1955), NACA Tech. Note 3298.
SINHA, R., ZAKKAY, V., and ERDOS, J. (1971). *AIAA J.*, **9**, 2263.
SMITH, F. T. (1969). In *Kinetic processes in gases and plasmas* (ed. A. R. Hochstim), p. 257. Academic Press, New York.
SPRINGER, G. S. (1971). In *Advances in heat transfer*. **7**, p. 163. Academic Press, New York.
STURTEVANT, B. and STEINHILPER, E. A. (1974). In *Rarefied gas dynamics* (ed. K. Karamcheti), p. 159. Academic Press, New York.
TALBOT, L. and SHERMAN, F. S. (1959). NASA Memo 12-14-58W.
TANNEHILL, J. C., MOHLING, R. A., and RAKICH, J. V. (1974). *AIAA J.*, **12**, 129.
TOENNIES, J. P. (1974). In *Rarefied gas dynamics* (eds. M. Becker and M. Fiebig), p. A1, DEVLR Press, Porz-Wahn, Federal Republic of Germany.
TUER, T. W. and SPRINGER, G. S. (1973). *Int. J. Computers and Fluids*, **1**, 399.
UHLENBECK, G. E. (1974). In *Rarefied gas dynamics* (ed. K. Karamcheti) p. 3. Academic Press, New York.
VICTORIA, K. J. and WIDHOPF, C. F. (1973). In *International conference on numerical methods in fluid mechanics*, p. 254. Springer-Verlag, Berlin.
VINCENTI, W. G. and KRUGER, C. H. (1965). *Introduction to physical gas dynamics*. Wiley, New York.
VOGENITZ, F. W., BIRD, G. A., BROADWELL, J. E., and RUNGALDIER, H. (1968). *AIAA J.*, **6**, 2388.
——, BROADWELL, J. E., and BIRD, G. A. (1970). *AIAA J.*, **8**, 504.
WALDMANN, L. Z. (1959). *Z. Naturforsch.*, **A14**, 589.
WENAAS, E. P. (1971). *J. chem. Phys.*, **54**, 376.
YANG, H. T. and LEES, L. (1960). In *Rarefied gas dynamics* (ed. F. M. Devienne), p. 201. Pergamon, London.
YEN, S. M. (1973). *Int. J. Computers and Fluids*, **1**, 367.

AUTHOR INDEX

SUBJECT INDEX